A PRACTICAL GUIDE TO MONOCLONAL ANTIBODIES

A PRACTICAL GUIDE TO MONOCLONAL ANTIBODIES

J. Eryl Liddell

and

A. Cryer

Department of Biochemistry, University of Wales College of Cardiff, UK

JOHN WILEY & SONS

Chichester · New York · Brisbane · Toronto · Singapore

Other Wiley Editorial Offices

John Wiley & Sons, Inc., 605 Third Avenue,
New York, NY 10158–0012, USA

Jacaranda Wiley Ltd, G.P.O. Box 859, Brisbane,
Queensland 4001, Australia

John Wiley & Sons (Canada) Ltd, 22 Worcester Road,
Rexdale, Ontario M9W 1L1, Canada

John Wiley & Sons (SEA) Pte Ltd, 37 Jalan Pemimpin   05–04,
Block B, Union Industrial Building, Singapore 2057

Library of Congress Cataloging-in-Publication Data:

Liddell, J. Eryl.
    A practical guide to monoclonal antibodies / by J. Eryl Liddell
and A. Cryer.
        p.   cm.
    Includes bibliographical references.
    Includes index.
    ISBN 0 471 92905 0 (spiral bound)
    1. Monoclonal antibodies. 2. Hybridomas. I. Cryer, Anthony.
II. Title.
    [DNLM: 1. Antibodies, Monoclonal. 2. Biotechnology—methods. QW
575 L712p]
QR186.85.L53   1991
616.07'93—dc20
DNLM/DLC
for Library of Congress   90-13056
                                                                   CIP

British Library Cataloguing in Publication Data:

Liddell, J. Eryl
    A practical guide to monoclonal antibodies.
    1. Monoclonal antibodies
    I. Title II. Cryer, Anthony
    574.293

    ISBN 0 471 92905 0

Typeset by Dobbie Typesetting Limited, Tavistock, Devon
Printed in Great Britain by Courier International, Tiptree, Colchester

Contents

Preface

This book has evolved from a course manual written for one-week short post-experience courses in Monoclonal Antibody Technology held since 1984 at the Department of Biochemistry, University of Wales College of Cardiff, UK. The book is designed, as are the courses, to provide all the information required by a competent scientist to produce monoclonal antibodies and to prepare them for use in a given application. The procedures are dealt with in chronological order starting with basic tissue culture techniques if required, immunisation strategies and screening test design, followed by production of hybridoma cell lines and basic antibody characterisation, purification and labelling. The chapters contain explanatory text on each step with comparative analysis of methods where appropriate. All necessary experimental protocols are presented in a self-contained format that should be easy to follow at the laboratory bench. Alternative protocols are provided where particularly relevant but for others that are not included in full, source references are given. The concluding chapter presents an overview of the current status of human hybridoma production and antibody engineering using techniques of molecular biology.

In relation to the production of this book we are grateful to Mr Guy Pitt of this Department for most of the photographs. All other photographs provided by manufacturers are credited individually. We would also like to thank all the numerous people who have taught and demonstrated on the courses over the years, contributors from industry and especially course delegates whose feedback is essential in helping us to provide the level of tuition demanded in this continually evolving technology.

J. E. Liddell
A. Cryer

Cardiff, August 1990

Glossary

Adaptive immunity   An acquired immunity which is specific for the inducing antigen. It is enhanced by repeated exposure to antigen.

Adjuvant   Compound administered with immunogen in order to enhance the immune response.

Adjuvant peptide   Usually refers to muramyl dipeptide which is the smallest active part extractable from the cell wall of BCG.

Allotypes   Structural variants of immunoglobulin that are characteristic of particular individuals of a species.

Antibody   Antigen exposure induces the production of this class of serum protcins. Antigen and antibody bind in a complementary and specific fashion.

Antibody affinity   Measure of the strength of the interaction between an antibody-combining site and an antigenic determinant.

Antibody avidity   Measure of antigen–antibody binding strength incorporating aspects of affinity and valency.

Antibody titre   The relative concentration of antibody present in a sample of serum or other antibody-containing fluid.

Antibody valency   Number of antigen-binding sites on any immunoglobulin molecule (e.g. $IgG = 2$, $IgM = 10$).

Antigen   The molecular entity which has the capacity to react with or bind selectively to an immunoglobulin.

Antigen-presenting cells   Cells which contribute to the processing, presentation and transport of antigens to the cells of the immune system.

Antigen receptors   On B-cells these are membrane immunoglobulins; on T-cells they are dimers of so-called a and b chains.

Antigenic determinant   The part of an antigen to which the antibody binds; an antigen may have many different determinants.

Antiserum   A serum collected following immunisation.

Ascites/ascitic fluid   Fluid product of intraperitoneal tumour growth. Induced in rodents after administration of Pristane and hybridoma cells.

B-cells   See B-lymphocytes.

B-cell mitogen   See mitogen, e.g. endotoxin, pokeweed mitogen.

Bispecific monoclonal antibodies   Antibodies in which each of the two antibody-combining sites are of different specificity. They are derived from hybrid hybridomas.

B-lymphocytes   Recognise 'foreign' material and produce antibodies.

Capping   The process of aggregation of cell surface molecules on the cell membrane, usually in response to antibody binding.

C genes   Gene segments encoding the constant regions of immunoglobulin heavy and light chains.

Chimeric antibodies   Genetically engineered antibodies in which parts of the recombinant immunoglobulin molecule are derived from different animal species.

Clonal selection   The process of antigen-induced selection, proliferation and maturation of specific lymphocyte clones.

Clone   A group of cells derived from a single cell and therefore exhibiting genetic identity.

Cloning   The separation of monoclonal cell lines from an originally mixed cell population.

Complement   Group of plasma proteins which interact with antibodies to mediate inflammation and the opsonisation of antigenic particles and destroys cells and pathogens.

Complementarity-determining region (CDR)   Areas of antibody molecule responsible for binding to antigen determinant.

Confluence   Cell culture density in which cells are in close proximity throughout the culture. Myelomas and hybridomas will be growing slowly and will die if not subcultured.

Constant region   Relatively invariant regions of immunoglobulin heavy and light chains responsible for effector functions of the antibody.

Cross-reaction   Binding of an antibody to a molecule not present in the immunisation mixture.

Determinant   See antigenic determinant.

D genes   Gene segments between V and J genes in immunoglobulin heavy-chain genes.

Domain   Globular region in a polypeptide chain as found in immunoglobulin molecules.

Electrophoresis   The separation of the components of a mixture according to their charge, in an electric field.

Enzyme-linked immunosorbent assay (ELISA)   Chromogenic immunoassay in which the analyte is measured with an enzyme-conjugated antibody and the enzyme substrate.

Epitope   See antigenic determinant.

Fab fragment   Monovalent antibody fragment containing antigen-binding site resulting from papain digestion of immunoglobulin.

F(ab')$_2$ fragment   Divalent antibody fragment containing antigen-binding sites resulting from pepsin digestion of immunoglobulin.

Fc fragment   Fragment of immunoglobulin heavy chain resulting from papain digestion of immunoglobulin.

Fc receptor   Molecule which binds Fc part of immunoglobulin.

HAT medium/selection   Culture medium containing hypoxanthine, aminopterin and thymidine, in which myeloma cells deficient in the enzyme HGPRT cannot survive.

Hybridomas with one 'normal' parent inherit the ability to make this enzyme and so are 'selected' out of a cell mixture by their ability to survive.

Heavy chains A pair of longer peptide chains which, with a pair of shorter light chains, form the immunoglobulin molecule.

Hinge Area of immunoglobulin heavy chain between the Fc and Fab regions which allows the complete immunoglobulin greater flexibility.

Histocompatibility The sharing of certain genes between recipient and donor individuals which will determine the acceptance of a transplant/graft.

Hollow fibre reactor Device for intensive culture of hybridomas and therefore antibody propagation. Composed of a bundle of hollow plastic fibres contained in a chamber. Hybridomas grow in the extracapillary space and are fed by culture medium pumped through the lumen of the fibres. The fibre pore size allows nutrients through to the cell compartment but antibody is prevented from leaking into the reservoir.

Hybrid hybridoma A hybridoma cell line derived from fusions between two different hybridoma lines, or one hybridoma and an immune lymphocyte. Bispecific antibodies can be secreted.

Hybridoma A cell line produced in vitro by the fusion of a malignant and normal cell (i.e. myeloma and lymphocyte).

Hypervariable region The most variable regions within the variable region of immunoglobulin. They contribute to the antigen-binding site.

Idiotope A single antigenic determinant in antibody variable regions.

Idiotype The antigenic characteristic of the variable regions of antibody.

Immunoadsorbent A solid-phase complex of antibody and carrier used in certain immunoassays or immunopurification to bind to antigen.

Immunogen Substance capable of inducing an immune response.

Immunogenicity The degree to which an antigen can elicit an immune response.

Immunoglobulin See antibody.

Immunoradiometric assay (IRMA) Assay in which the substance to be measured (analyte) is directly proportional to the binding of radiolabelled antibody.

Isotype Structural variant of immunoglobulin within which molecules have similar or identical constant regions. Also called class or subclass, i.e. IgM, IgG_1, IgG_{2a} etc.

J genes Genes encoding for part of the variable regions of both immunoglobulin heavy and light chains.

Kappa chain One of the immunoglobulin light-chain isotypes.

Lambda chain One of the immunoglobulin light-chain isotypes.

Light chains A pair of shorter peptide chains which, with a pair of longer heavy chains, form the immunoglobulin molecule.

Liposome Artificially generated small lipid membrane vesicle, that can be used as an adjuvant.

Log phase A period of exponential cell growth in which the cells will be at their most homogeneous and of high viability. Myelomas for fusion should be in this phase.

Lymphocytes A group of blood cells involved in mounting and controlling the immune response. Subgroups include B (bone marrow derived) and T (thymus derived) cells.

Lymphoid Description of the (primary) tissue from which lymphocytes arise (bone marrow, thymus) and the (secondary) tissues (lymph nodes, spleen) where lymphocytes reside.

Lymphokines Non-immunoglobulin products of lymphocytes which help intercellular communication in the immune system.

Macrophages Large, long-lived phagocytic cells.

xiiGLOSSARY

Mast cells Cells able to release inflammatory mediators in response to a range of stimuli.

Memory cells Lymphocytes that maintain an immune memory following a primary response.

Mitogens Molecules able to produce differentiation and division of cells.

Monoclonal Cells or their secreted antibody derived from one clone of B-lymphocytes.

Myeloma (Plasmacytoma) Tumour of malignant plasma cells.

Nude mouse/rat A genetically athymic animal which also carries a closely linked gene causing hairlessness.

Plasma cells Terminally differentiated B-lymphocytes producing antibody.

Polyclonal Cells or antibodies derived from many clones of B-lymphocytes, i.e. as in antiserum.

Primary immune response Response elicited following first exposure to an antigenic challenge (immunisation).

Radioimmunoassay (RIA) Assay based on competition of analyte with radiolabelled analogue of the analyte for a limited number of antibody sites. The concentration of the analyte is therefore inversely proportional to the amount of label bound to the antibody.

Recombination The rearrangement of genetic information during meiosis. Can be manipulated to bring together and join otherwise separated gene sequences.

Secondary immune response Response elicited following second (or subsequent) exposure to an antigenic challenge (immunisation).

Single domain antibodies (dABs) Recombinant V_H domains expressed in *E. coli* that possess binding affinities with antigen comparable to whole antibodies.

Somatic mutation Involves base changes in the DNA and consequent point mutations in the encoded protein, that occur during the life of the cell.

Subculture Or expansion, when cells are diluted in fresh medium and transferred to more or larger culture vessels in order to maintain the high-viability log phase of growth.

T-lymphocytes Have a number of roles in facilitating and controlling the immune response, including functions relating to the destruction of infective agents.

Transfection Process of artificially introducing genetic material into a cell.

Transformation (cellular) Process whereby normal control of cell division is lost.

Variable region Area of heavy and light chain of immunoglobulin molecule with a relatively high degree of amino acid sequence variation, within which lies the antigen-combining site.

V genes Genes encoding for the variable regions of immunoglobulin heavy and light chains.

Abbreviations

Ab	antibody
ABEN	7-N-(4-aminobutyl-N-ethyl)-napthalene-1,2-dicarboxylic acid hydrazide
ABTS	2,2'-azino-di-(3-ethyl-benzthiazoline-sulphonate-6)
AE	4-(2-succinimidyloxycarbonylethyl)-phenyl-10-methylacridinium-9-carboxylate fluorosulphonate
AEC	3-amino-9-ethylcarbazole
AET	2-aminoethyl isothiouronium bromide
8-AG	8-azaguanine
Ag	antigen
AP	alkaline phosphatase
5-AS	5-aminosalicylic acid
BCIP	bromochloroindolyl phosphate
β-G	β-galactosidase
BNHS	biotinyl-N-hydroxysuccinimide ester
BSA	bovine serum albumin
BSF-2	B-cell stimulatory factor
BX-NHS	Biotinyl-ϵ-aminocaproic acid N-hydroxysuccinimide ester
CDI	carbodiimide
CDR	complementarity-determining region
CFA	complete Freund's adjuvant
CHM-NHS	4-(N-maleimidoethyl)-cyclohexane-1-carboxylic acid N-hydroxy-succinimide ester
μCi	microcurie
4-CN	4-chloro-1-naphthol
Con A	concanavalin A
cpm	counts per minute
CTL	cytotoxic T-cells

DAB	3,3'-diaminobenzidine
dABS	single domain antibodies
DBM	diazobenzyloxymethyl
DEAE	diethylaminoethyl
DMEM	Dulbecco's minimal essential medium
DMF	dimethylformamide
DMSO	dimethylsulphoxide
DNA	deoxyribonucleic acid
EBNA	EBV-specific nuclear antigen
EBV	Epstein–Barr virus
E. coli	*Escherichia coli*
EDTA	ethylene diamine-tetraacetic acid
ELISA	enzyme-linked immunosorbent assay
Fab	see Glossary
F(ab')$_2$	see Glossary
FACS	fluorescence activated cell sorting
FITC	fluorescein isothiocyanate
FPLC	fast protein liquid chromatography
FBS	foetal bovine serum
Fc	see glossary
FCS	foetal calf serum
g	force of gravity
g	gram
μg	microgram
GA	glutaraldehyde
HAT	hypoxanthine–aminopterin–thymidine
HEPA	high-efficiency particle filter
HEPES	*N*-2-hydroxyethylpiperazine-*N'*-ethanesulphonic acid
HGF	hybridoma growth factor
HGPRT	hypoxanthine guanine phosphoribosyl transferase
h/h	human/human
HPLC	high performance liquid chromatography
HRP	horseradish peroxidase
HT	hypoxanthine thymidine
i.d.	intradermal
IFA	incomplete Freund's adjuvant
Ig	immunoglobulin
IL-6	interleukin-6
i.m.	intramuscular
i.p.	intraperitoneal
IRMA	immunoradiometric assay
i.v.	intravenous
$\varkappa$	kappa
K	affinity constant
KLH	keyhole limpet haemocyanin
λ	lambda
μl	microlitre
LCL	lymphoblastoid cell lines
MBS	*m*-maleimidobenzoyl-*N*-hydroxysuccinimide ester

μm	micrometre
2-ME	2-mercaptoethanol
M	molar
μM	micromolar
MDP	muramyl dipeptide
m/h	mouse/human
MHC	major histocompatibility complex
MHz	megahertz
ml	millilitre
mm	millimetre
mM	millimolar
6-MPDR	6-methylpurine deoxyriboside
MW	molecular weight
NBT	nitroblue tetrazolium
nm	nanometre
p-NPP	p-nitrophenyl phosphate
OD	optical density
ONPG	o-nitrophenyl-β-D-galactosidase
OPD	o-phenylenediamine
OPDM	N,N'-o-phenylenedimaleimide
PBS	phosphate-buffered saline
PEG	polyethylene glycol
pg	picogram
PHA	phytohaemagglutinin
PPLO	pleural pneumonia-like organism
R	unit of radiation
RER	rough endoplasmic reticulum
RIA	radioimmunoassay
RNA	ribonucleic acid
rpm	revolutions per minute
RPMI	Roswell Park Memorial Institute
SAS	saturated ammonium sulphate
s.c.	subcutaneous
SDS	sodium dodecyl sulphate
SDS-PAGE	sodium dodecyl sulphate–polyacrylamide gel electrophoresis
SMPB	succinimidyl 4-(p-maleimidophenyl)-butyrate
SPDP	N-succinimidyl 3-(2-pyridylthio)-propionate
SRBC	sheep red blood cells
6-TG	6-thioguanine
TK	thymidine kinase
TMB	3,3,5,5'-tetramethylbenzidine
Tris	Tris-(hydroxymethyl)-aminomethane
TRITC	tetramethyl rhodamine isothiocyanate
U	units
UV	ultraviolet
v/v	volume for volume
w/v	weight for volume
w/w	weight for weight
XGPRT	xanthine guanine phosphoribosyl transferase

Protocols

Chapter 1

Overview of hybridoma technology

1.1 INTRODUCTION

Monoclonal antibodies have now been part of the rich tapestry of biological knowledge for over a decade. Although enormous advances in our understanding of these molecules have occurred in that time, their strategic application still constitutes a major growth area in biological, biotechnological and clinical science.

In 1986, Kohler and Milstein were awarded the Nobel Prize, in recognition of the importance of their contribution to the development of means for the production of monoclonal antibodies (Kohler & Milstein, 1975; Milstein, 1986). But their true prize must be the realisation that their pioneering work has led to an explosive improvement in our understanding of immunology and has produced new possibilities for the investigation, diagnosis and treatment of many hitherto poorly understood diseases.

Before entering into a fuller description of the constituent parts of the strategy that can be thought of as the 'monoclonal antibody approach' it is necessary to cover some essential theoretical aspects of the process first. Additionally, it may be useful for some readers to have an introductory overview of the strategy in order to provide a contextual framework, into which the practical aspects of what follows can be placed.

1.2 BACKGROUND

The immune system of higher animals has as its essential function the provision of a means of defence against infection. In essence, this function is achieved by a combination of what are referred to as innate and adaptive mechanisms. The protection that the immune system provides, as far as the current considerations are concerned, relates mainly to that part of the response which is adaptive in nature. This adaptive response is characterised by the recognition and elimination by the body of material that may be described as non-self (be this an infective agent or other foreign material). The immune response is a co-ordinated set of processes involving a variety of specific cell types, of which the most important are the lymphocytes. Lymphocytes are of two major types, being involved in either antibody production, the B-lymphocytes, or in the control of cell-based destructive mechanisms and in regulating the intensity of the overall response (T-lymphocytes).

Antibodies, as a specialised protein class produced by B-lymphocytes, are able to interact specifically with foreign materials and to neutralise any potentially harmful properties that they may have and, in addition, facilitate their eventual elimination. Although any particular antibody-producing lymphocyte is only able to produce a single type of antibody molecule, the overall response to challenge that is elicited in vivo depends on many cells. Each cell that is involved will have an individual origin, and will contribute a different antibody molecule species to the complex and concerted protective response.

Once stimulated in vivo, B-lymphocytes undergo processes of proliferation, differentiation and maturation such that many identical cells capable of producing antibody of a particular nature arise from a single or small number of progenitor cells. Such collections of cells, by definition, will be said to be monoclonal in origin.

The response registered in the serum of an individual will be made up of the products of a number of such monoclonal cell pools and will, therefore, be polyclonal in origin.

The objective of the monoclonal strategy for the production of useful antibodies revolves, therefore, around the selection of antibody-producing cells of monoclonal origin from the multiplicity of such cells present in vivo following the administration of material that is recognisably non-self, the selection being based on the detection of the single molecular species of antibody that a particular cell clone produces. Thus, not only must antibody-producing cells be maintained outside the body of the donor, but they must also be treated in such a way that selection procedures can be instituted and the cells of choice can have their monoclonality proven and their antibody product harvested in useful quantities before the successful production of monoclonal antibody can be claimed.

1.3 THE CELL PROBLEM AND ITS SOLUTION

The basic problem, which had hitherto inhibited progress in this area, was the extreme difficulty in culturing antibody-producing cells outside the body for any experimentally or operationally meaningful period of time. The solution to this problem, like many others, came with the application of procedures which previously had been restricted in their use to a separate, but not totally unrelated, area of endeavour. Hybrid cells produced by the induced fusion of dissimilar cell types had for many years been a recognised tool in the study of cellular genetics. What Kohler and Milstein were able to show was that antibody-producing cells could be hybridised. What is more they showed that this could be done with cell partners that conferred on the hybrid the extremely useful property of potentially infinite life in vitro without compromising the capacity of the hybrid to synthesise and secrete antibody. With this basic capability to hand, they were then able to devise means whereby cells producing a chosen antibody of predetermined specificity could be selected from the multiplicity of those present in vivo and these selected cells could then be propagated successfully in vitro.

1.4 THE STRATEGY

The overall strategy for the production of monoclonal antibodies is shown in Fig. 1.4. As can be seen, this involves the administration of an antigen-containing preparation to an appropriate recipient animal. The treated animal then becomes the source of sensitised antibody-producing cells. These cells are then mixed with a specially selected immortal cell line and the mixture exposed to an agent which promotes the fusion of cells. Under these circumstances fusion between cells takes place in a random manner and a fused cell mixture together with unfused cells of each type is the product. The immortal cell partner for the antibody-producing cell is the myeloma cell, itself a cancerous cell derived from the immune system. The myeloma cell lines that are used are specially chosen such that, by the use of specific selection media, the only cells to persist in culture from the fusion mixture are those that are hybrids between cells derived from the immunised donor and myeloma cells, i.e. hybridomas.

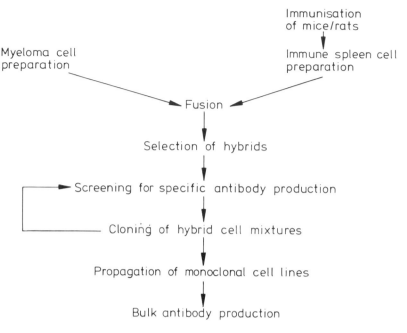

Fig. 1.4

The other possible combinations of fused cells are not able to survive the selection pressure that is applied.

At this stage, the hybrids that survive will still be a very mixed population, not all of which will be able to synthesise antibody. Furthermore, of those that can synthesise antibody, only a few may exist that are able to produce antibody which has the predetermined specificity towards the chosen antigen.

There is a need, therefore, to apply two further procedures in order to bring the strategy to fruition. Firstly, a screening test is required which will help to determine whether any particular cell culture is capable of synthesising antibody and, more particularly, antibody with the desired specific reactivity. Thus, following fusion, the cellular products are distributed among many small growth containers and allowed to establish themselves (or otherwise) under the specific media conditions that support only hybridomas. Then the growth media in which the cells have grown is tested for the presence of specific antibody reactivity. Those cultures that are positive in this regard are then chosen for further study. It is noteworthy that the number of such cell cultures that may be considered positive may range from a very small proportion to a quite significant proportion of the total number present. This will depend on the nature and behaviour of the immunising material and the specificity of the screening test employed. However, those cultures that are chosen will, in all likelihood, contain not only cells secreting the antibody which has the specificity of choice but also a variable number of other cells that may be secreting other irrelevant antibodies.

The next stage of the strategy must therefore be separation of the hybridomas capable of synthesising and secreting the antibody of choice from all the other irrelevant ones that are present. This is achieved by the process of cloning. It involves the repeated growth of cell colonies from very low original numbers until it can be adduced with certainty that the colony has been derived from a single cell. By definition, such a cell culture will be monoclonal in origin and, if it retains the capacity to synthesise antibody, then those antibodies will be monoclonal antibodies in that all the cells present will be synthesising an identical antibody gene product. Such monoclonal antibody preparations will exhibit reactivity with a single antigenic determinant. It is this property of specificity that makes monoclonal antibodies invaluable in such a wide range of preparative, analytical, diagnostic and therapeutic situations. Since, no doubt, most readers may already have such applications in mind, this book will not concern itself with any attempt to catalogue them, but will restrict itself to the more practical aspects of monoclonal antibody preparation.

As a preliminary to the specific considerations of the techniques of monoclonal antibody production the next chapter deals with cell culture in general. This should enable those who are not familiar with this set of procedures to assess what is needed for the establishment and operation of a well-founded cell culture laboratory and to acquire these basic but essential skills.

Chapter 2

Tissue culture techniques

2.1 GENERAL CONSIDERATIONS

This chapter deals with the basic requirements, in terms of equipment and practical technique, necessary for the maintenance of cells in culture. It is aimed at those who have no previous experience of tissue culture. Specific techniques related to the production of hybridomas will be covered in subsequent chapters although the requirements of a monoclonal antibody production laboratory will be indicated here.

There are three essential pieces of equipment required for growing cells: an incubator in which the cells can be kept at an appropriate temperature for growth, an enclosure supplied with filtered air in which the cell containers can be opened and the cells they contain treated and a microscope for the observation of the cells. In addition, the cells must be provided with a nutritive medium especially developed to support the growth of particular cell types and the container in which they are kept must be compatible with the variety of cells involved. Although these requirements will be dealt with in detail below, with the special techniques necessary for successful maintenance of cells in culture, those readers without any experience of basic cell culture may find other valuable information in the many manuals of cell culture techniques that are available (e.g. Freshney, 1987; Paul, 1975).

The conditions for the maintenance of eukaryotic cells in culture are also ideal for the growth of contaminating micro-organisms. The presence of any such contaminating organisms is, of course, to be avoided at all costs. The practice of proper aseptic technique requires that all the reagents and containers etc. that come into direct contact with cells must be sterile. Although the introduction of sterile disposable plasticware has simplified the situation considerably, efficient organisation of both the culture area and the stock control of materials is essential for successful, contamination-free work. As a starting point for this, good laboratory practice should be exercised at all times and constant vigilance, with regard to the eradication of potential sources of infection, should be adopted.

2.2 EQUIPMENT REQUIREMENTS

2.2.1 The clean area

In an ideal situation, tissue culture would be carried out in rooms having their own filtered air supply. Such rooms would have positive air pressure within them, be subject to strictly restricted access and have a number of other special requirements designed to act as barriers to contamination entering from outside. Although there are a number of companies specialising in the supply of such clean rooms, who can advise on individual requirements (see Appendix B), in many practical circumstances this level of provision is neither convenient, achievable nor affordable.

Fortunately, the introduction of laminar flow cabinets/hoods has made the creation of an adaptable and acceptable tissue culture laboratory very much simpler and cost effective. Indeed, an ordinary laboratory can be adapted initially for tissue culture use, merely by the installation of a laminar flow hood, and the facility can be virtually completed by the provision of ready access to an incubator and a microscope.

However, various aspects of such an undertaking will benefit from careful forward planning. Firstly, when a conversion is to take place, the location of the culture laboratory must be given careful consideration. For instance, if a choice of locations is available, the laboratory should not be situated near to any microbiological work area or animal house and should not open directly onto corridors with heavy pedestrian use. Any windows in the laboratory should be permanently sealed to prevent dust penetration.

Secondly, all the necessary manipulations involving cells should be confined strictly to the laboratory, with the exception of animal dissections, which should never be performed in the culture area. A warm room and/or incubators should be incorporated within the tissue culture laboratory itself, conveniently close to the laminar flow hood and microscope. The laboratory should be in close association with the necessary storage, preparation and wash-up areas with non-tissue culture activities within the overall area being avoided or, at the very least, strongly discouraged. The designated laboratory should also be kept totally free of unnecessary pieces of equipment. In order to facilitate cleaning, the floor and bench surfaces in the tissue culture laboratory should be made of impervious, continuously laid materials and the walls should be painted with waterproof, oil-based gloss paint. The joints between walls, floor and ceilings should be suitably sealed, such that they may be kept scrupulously clean.

Because of the long-term culture involved in monoclonal antibody production, it is preferable to segregate all these cell-based activities from other culture work, especially that which might involve the establishment of primary cell lines. This means that primary cell lines should never be kept in the same incubator or dealt with in the same laminar flow hood as hybridoma cells. If resources allow, it should be a desired objective that a separate hood and incubator be made available for each operator. Whatever the level of physical provision that is created, always keep in mind that valuable and productive hybridoma cell lines can be lost to contamination or other mishap much more easily than they can be replaced.

2.2.2 Laminar air flow cabinets

All the necessary manipulations of culture medium and cells should be carried out in a sterile environment, in order to help prevent contamination by airborne organisms. Such an environment is provided by laminar air flow cabinets or hoods. These operate on the principle that an enforced flow of filtered air is directed over the work area. The filters that are used are high-efficiency particle filters (HEPAs) capable of removing from air particulate matter that is greater than $0.3\,\mu$m in diameter. Thus most, but not all, bacteria and fungal spores etc. that are normally present in air will be removed. In addition any aerosols generated during cell handling will tend to be removed rapidly from the work area by the air flow.

The effectiveness of laminar flow cabinets depends on their individual performance capability, on the location of the cabinet within the laboratory and on the operational practices that are adopted. Thus, any disturbance of the air flow within the cabinet, which may be caused by a variety of factors such as its proximity to doorways, the passage of pedestrian traffic, the operation of external fans, the presence of too much equipment within the cabinet, thermal currents or the excessive interruption of the air flow by the operator, will impair the efficiency of the cabinet and thereby put the work at risk. The air flow in the cabinet should, in any case, be measured regularly with an anemometer to ensure that it is both steady and up to specification. No decline in performance should be expected from a cabinet that is serviced regularly.

Cabinets are available in a number of configurations. They vary in size and have either vertical or horizontal air flow. Examples are shown in Fig. 2.2.2. Some cabinets are also fitted with UV lamps that help to maintain the sterility of the environment. Care must always be exercised with such fittings and they must always be switched off

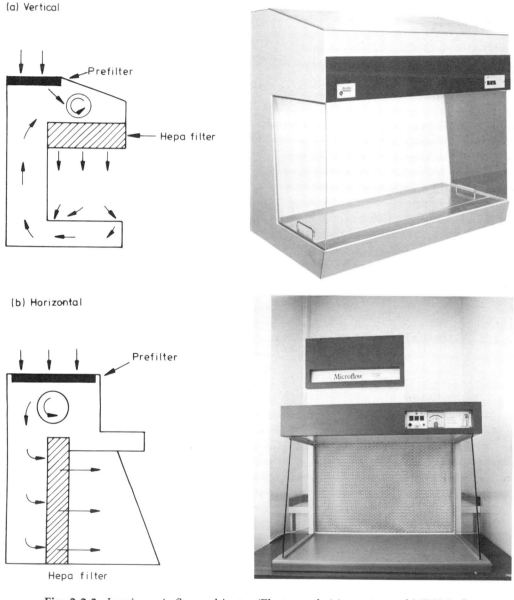

(a) Vertical

Prefilter

Hepa filter

(b) Horizontal

Prefilter

Hepa filter

Microflow

Fig. 2.2.2 Laminar air flow cabinets. (Photograph (a) courtesy of MDH Ltd)

when the cabinet is in use or during the working day. The necessary notices warning of the presence of any UV sources should also be displayed.

Depending on their design, cabinets can not only protect the work from contamination but can also offer various levels of protection to the operator. A classification based on the level of protection afforded by particular pieces of equipment has been adopted and encompasses Classes I–III. In particular, Class I cabinets are adequate for that level of basic cell culture which does not involve cells or organisms of potential hazard to the operator. They are not suitable, for example, when cells of human origin or cells that are virally infected are being manipulated. Cabinets with horizontal air flow which may not be classified and are sometimes referred to as 'work stations', in which virtually all the filtered air is vented directly at the operator, should only be used for basic cell manipulations where it is certain that there is no risk to the operator.

Class II cabinets offer extra protection to the operator by recirculating the air that has passed over the work through the filtration system such that no unfiltered air escapes from the enclosure. Some of these cabinets also have facilities which allow for the venting of recirculated, filtered air to the outside. Class III cabinets are also available for particularly hazardous work where, for example, pathogens are involved. Advice on which class of cabinet is required for any particular application can be obtained from the Health and Safety Executive in the UK.

2.2.3 Incubators

Most cells from endothermic animals need to be kept at a temperature of 37°C for optimal growth. Incubators, used to create such conditions, should be capable of providing temperatures that are both constant and uniformly distributed throughout the incubator enclosure. To achieve this, most incubators have a thermostatically controlled water jacket as an integral part of their construction. Because of the weight of the water in such jackets, incubators of this design need to be situated on the floor

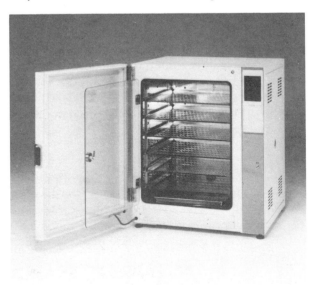

Fig. 2.2.3 CO_2 incubator. (Photograph courtesy of New Brunswick Ltd)

(preferably on a movable platform to facilitate cleaning) or on reinforced benching. It is easy, when choosing an incubator for the first time, to underestimate the size that will be required. Since requirements inevitably grow, one should purchase the largest incubator that is affordable, or preferably two, to insure against the breakdowns and infections which will almost certainly occur at infrequent, yet inevitably inconvenient, times.

For many applications in general cell culture, temperature control is all that is required from an incubator. However, in the case of monoclonal antibody production, hybridoma cells prefer conditions in which bicarbonate acts as a buffering ion, helping to maintain the pH of the culture medium. In order for this buffering system to work, the cultures and their medium must be allowed to equilibrate with an atmosphere enriched in CO_2. Thus, the incubators required should be capable of maintaining an internal atmosphere of 5% CO_2 in air. The CO_2 is supplied from cylinders of compressed gas connected through a CO_2 sensor to the atmosphere inside the incubator. In addition, the growth containers used should not be gas-tight and the incubator should be kept humidified to prevent the cultures drying out. The humidity within the incubator is kept high by the insertion of a shallow tray of sterile water. If the cultures are to be kept for any length of time, it is also important for the CO_2 control to be automatic, preferably with a cylinder change-over device. Then, if the gas cylinder empties unexpectedly, the interior atmosphere of the incubator is protected by the automatic switching to a new gas supply. CO_2 cylinders without a dipstick should always be used so that frozen CO_2 does not accumulate in the control valve, thus blocking the supply of gas and damaging the valve.

2.2.4 Microscope

Daily microscopical observation of cells in culture is essential for the assessment of cell growth and for the detection of possible infection at the earliest stages. Cultured cells, being alive and unstained, cannot be viewed effectively with simple transmitted light. Their observation requires a microscope fitted with phase-contrast optics. There are many relatively simple examples of such microscopes on the market. However, the chosen microscope is the one piece of equipment where extra expense is well justified. A good microscope is also essential if photomicrography or fluorescence microscopy is envisaged (see later). For hybridoma work, a ×10 and a ×20 objective with ×10–15 eyepieces give sufficient magnification of 200–300, for both scanning and more detailed observation of cultures. The fitting of a mechanical (movable) rather than fixed stage on the microscope aids the viewing of the many separate cultures on multiwell plates.

For mycoplasma detection using Hoechst stain an upright fluorescent microscope with a particular combination of filters (i.e. excitation filter 340–380 nm; suppression filter 430 nm) will also be required. However, its use will be infrequent enough that its purchase will not be justified for this purpose alone, if funds are limited. Access to such facilities on a shared basis is usually possible in larger institutions where other biomedical work is being carried out.

Having chosen a microscope, it is essential to learn how to set it up and maintain it so that it can be used to its full potential. All users should be acquainted

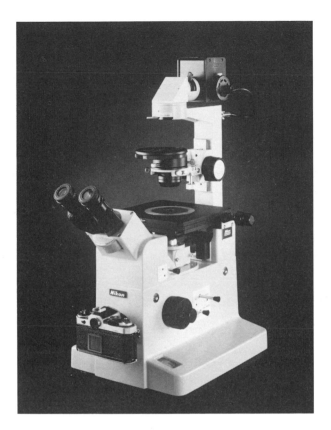

Fig. 2.2.4 Nikon Diaphot phase-contrast microscope. (Photograph courtesy of Nikon UK Ltd)

with the procedures described in the manual supplied with the microscope. However, specialist manufacturers are usually more than happy to demonstrate these procedures in detail.

2.2.5 Cell growth containers

Both cells and culture medium need to be kept in sterile containers. All of these used to be made of glass before the introduction of disposable plastic, which has eased the routine of tissue culture considerably, but has also increased the expense. However, in assessing costs, the time taken to wash and sterilise non-disposable items must be taken into consideration.

Glass is not usually used for growing cells, because of its relatively poor optical qualities, but it can be conveniently reused for the storage of medium after thorough washing, rinsing and autoclaving. One should ensure that the caps are capable of withstanding repeated autoclaving. Suitable bottles are manufactured by Duran.

There are many different forms of disposable plastic suitable for cell culture, ranging from multiwell dishes holding as little as 200 μl per well to roller bottles holding litres (Fig. 2.2.5a). Some are treated to encourage attachment of particular cell types, but all are supplied sterile and are of such quality as not to impair observation under the

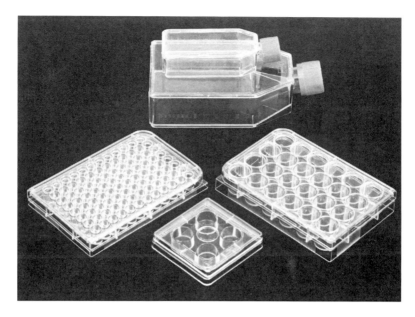

Fig. 2.2.5a Cell growth containers. (Various manufacturers; see Appendix B)

microscope. A range of most of these will be needed for hybridoma work, and are available from many manufacturers. The market for tissue culture materials is very competitive so the major difference between the best manufacturers is price. Discounts can usually be negotiated for bulk orders.

Large-scale production of cells or their secreted products is often impractical using the plastic containers described above. For this purpose, fermenters and more recently hollow fibre reactors have been developed. These are discussed in detail in the chapter dealing with the propagation of monoclonal antibodies.

In addition to growth containers, sterile centrifuge tubes will also be required, for washing cells and harvesting culture supernatant. The most useful sizes available hold 11–15 ml and 50 ml. Various universal containers and vials for storage of culture supernatant and cells are shown in Fig. 2.2.5b. A special quality of vial is available for storage of cells in liquid nitrogen which can withstand such low temperatures.

2.2.6 Pipetting devices

The transfer of medium and cells between containers should be effected in such a way that contact with non-sterile surfaces is avoided. In addition, exposure to external air, even if in a laminar flow hood, should be kept to the absolute minimum. Therefore, on both these counts, the pouring of liquids from one container to another is inadvisable.

The preferred means of transferring liquids is via some form of pipetting device, of which there is a large variety available. The choice of device will depend on the volumes of liquid to be transferred and the purpose of the transfer (see Table 2.2.6). Disposable items will add to the overall costs, but will save on the considerable amount of time taken up by washing and sterilising reusable units.

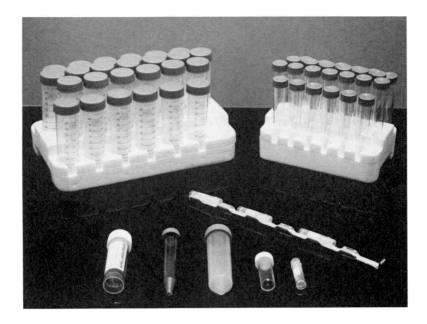

Fig. 2.2.5b Centrifuge tubes and storage vials for tissue culture. (Various manufacturers; see Appendix B)

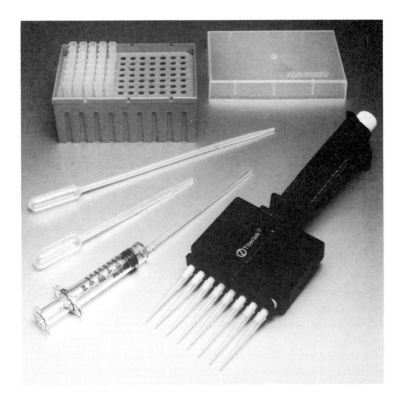

Fig. 2.2.6 Pipetting devices. Clockwise from top: sterile tips in autoclavable box; multichannel variable volume pipette; syringe with Kwill attached; plastic Pasteur pipettes (available sterile and individually wrapped)

Table 2.2.6

Volume of medium to be dispensed	Device
200 ml to 5 l	Automatic pump
1 ml to 50 ml	Syringes with Kwill*
1 ml to 10 ml	Glass or plastic pipettes with rubber bulb
<1 ml	Gilson/Eppendorf pipettes with sterile tips
0.5 ml to 3 ml	Disposable plastic pipettes with integral bulb
Multiple 5–200 μl	
From multiwell plates	Multichannel pipettes
To multiwell plates	Repeating pipettes or syringes with needles

*A Kwill (Avon Medical) is a plastic luer-fitting extension tube for a syringe.
See Appendix B for suppliers.

2.2.7 Autoclaves/ovens

Water, glassware, filters etc. can be sterilised by moist heat in an autoclave. There are various types and sizes of autoclave available commercially. These range from sophisticated, fully automatic devices to such simple pieces of equipment as the domestic pressure cooker. The principle of operation is the same for all of these, in that few organisms can withstand exposure to high pressure and temperatures for longer than one minute. The recommended conditions for complete sterilisation are exposure for at least 20 minutes at 10–15 lb/in^2 pressure and 120°C. The extra time allows steam to penetrate all parts of the autoclave and to superheat all of the contents. To ensure that sterilisation has been complete, it is useful to include indicators which are usually in the form of self-adhesive tapes. These change colour on exposure to high temperatures and may be affixed to glassware or used to secure packages of small items. With pressure cookers or older autoclaves it is necessary to replace all the internal air with steam before closure and allowing the pressure to rise. Then, after the recommended time, the pressure should be lowered gradually to prevent boiling of liquids. More modern autoclaves operate this procedure automatically through timers and electrically driven valves.

Solutions to be sterilised by autoclaving (N.B. not culture medium) should be contained in glass bottles. These should be stoppered with a loosely fitted screw cap covered in foil. On removing bottles from the autoclave these caps should be tightened immediately in order to ensure that the contents remain sterile.

Small objects can be wrapped in foil, gauze or nylon film bags prior to autoclaving. The orifices of any unsealed vessels such as flasks should be plugged with cotton wool covered with a foil cap. Glass pipettes, individually plugged with cotton wool, are usually autoclaved in a sealable metal container.

Glassware and dissecting instruments can also be sterilised by dry heat in standard laboratory ovens. However, for sterilisation to be complete, a longer exposure time and a higher temperature (i.e. 90 minutes at 160°C) are required to compensate for the poor conducting properties of dry air. Items can be wrapped in foil as for autoclaving or enclosed in appropriate metal containers. Temperature-sensitive indicator paint can be obtained for use as a guide to the completeness of dry heat sterilisation.

The methods described above cover most requirements. Plasticware is normally purchased already sterile for single use. However, if necessary, some plastic or metal items can be sterilised by alternative methods such as immersion in 70% alcohol for 30 minutes followed by drying in a laminar flow hood under UV light. Other effective methods of sterilisation such as exposure to ethylene oxide or gamma-irradiation are large-scale procedures usually unavailable to the general laboratory involved in cell culture and will not be considered further here.

2.2.8 Filtration devices

Culture medium and the various additives necessary to sustain living cells cannot be sterilised by heat. However, filtration through a 0.2-μm pore size filter will remove most micro-organisms likely to contaminate tissue culture. 0.45-μm filters or 'prefilters', whilst not capable of removing micro-organisms, can be useful for the removal of particulate matter prior to sterile filtration, thus increasing the volume that can be filtered before the filter becomes clogged. The expected volume throughput of a particular filter will depend mainly on the filter diameter or surface area, pore size, operating pressure, and viscosity of the solution. Recommended filter diameters for various volumes of culture medium without serum are given in Table 2.2.8.

Table 2.2.8

Volume to be filtered	Filter diameter
up to 50 ml (syringe volumes)	13 or 25 mm
100–400 ml	47 mm
1–5 litres	90 mm
5–20 litres	142 mm
>20 litres	293 mm

If reagents are bought as single strength sterile liquids they should not need to be filtered but they are usually more expensive than powder or concentrated stocks. The shelf life of powdered media is also longer than that of liquids. Any formulation that requires the addition of water will either have to be filtered or made up aseptically with autoclaved water. Filters can be obtained as disposable sterile units or reusable units, in which only the filter paper is replaced. Fig. 2.2.8 illustrates a selection of filters for different purposes.

2.2.9 Freezers

Fridges and freezers are, of course, standard storage equipment in ordinary laboratories, including those devoted to tissue culture. However, it is preferable to keep non-tissue culture items separate to minimise the possibility of cross-infection.

Culture medium is stored at 4°C and, whether bought as single strength liquid or prepared on site from concentrated liquid or powder, it will require appropriately large-capacity cold storage. A -20°C freezer is necessary for the storage of certain reagents to be added to culture medium, particularly foetal calf serum, which is normally ordered in large batches. However, most suppliers offer storage facilities for reserve batches.

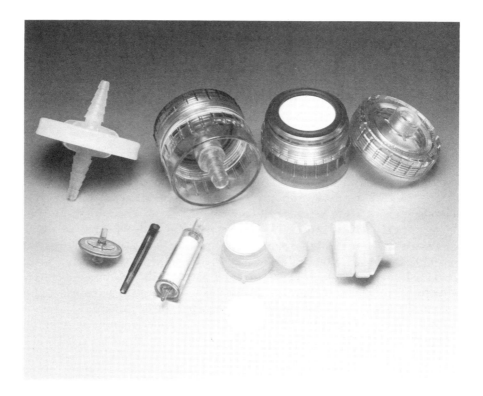

Fig. 2.2.8 Small-scale filtration devices. Top row, left to right: 47-mm diameter filters, (1) disposable, (2) reusable (closed), (3) reusable (open). Bottom row, left to right: (1–3) disposable, (4) reusable (open), (5) reusable (closed)

A $-70°C$ freezer is usually necessary as part of the process of freezing cells and for their short-term storage. Alternatively, for the cell-freezing process, there are devices on the market which can reduce temperature at a controlled rate (i.e. 1°C per min) before transfer of vials to liquid nitrogen storage.

The long-term preservation of cells requires storage at even lower temperatures, in liquid nitrogen. Management of liquid nitrogen storage is rather more complicated than that of electrical freezers. The cost of the storage containers and various alarm devices and refill facilities has to be balanced against the importance of having totally failsafe storage of what could be unique and invaluable cell lines. The smaller containers of 10–20-litre capacity will need refilling approximately every three weeks. Special deliveries of such relatively small volumes add to the expense and are not always reliable. The larger containers are much more expensive, although they can be connected to storage tanks for automatic refilling. Valuable cell lines should, in any case, be stored in more than one container to guard against accidents. Another consideration might be to store a valuable cell line in one of the National Cell Banks (see Appendix C) that will store cells for a fee and make them available to other laboratories unless requested otherwise.

2.2.10 Centrifuge

The washing, concentrating, and pelleting of cells requires centrifugation at rates of approximately 1500 rpm or 400 g. A simple bench centrifuge within the tissue culture laboratory is adequate. Higher centrifugation speeds will be required for later processes such as ascitic fluid clarification but should be carried out in the general laboratory.

2.3 PROCEDURES

The previous section has described the equipment necessary for a tissue culture laboratory. The general procedures necessary to carry out tissue culture are described below. The prevention of contamination in cultures is paramount. Inadequate precautions and lapses in standards can easily ruin months of work and result in the irretrievable loss of valuable cell lines, particularly in long-term tissue culture such as monoclonal antibody production. Recommended routine procedures are described for laboratory cleaning and personal hygiene (Protocols 2.3.1 and 2.3.2). This is followed by methods for general tissue culture practice (Protocol 2.3.3), and for the washing and sterilisation of equipment (Protocol 2.3.4). There follows a section dealing with the requirements of culture media and their preparation in general. Specific recipes for media and buffers etc. are given in Appendix A. Finally there is a more detailed section on the most likely types of infection to be encountered and the use of specific antibiotics.

PROTOCOL 2.3.1: ROUTINE CLEANING OF THE LABORATORY

Unless specialist cleaning staff are available, it is preferable that the tissue culturist, who will appreciate the necessary standards, is personally responsible for cleaning the laboratory.

Daily tasks:

1. Mop the floor with detergent, preferably at the end of the working day. Use a mop reserved for use only in the tissue culture laboratory.

Weekly tasks:

1. Wipe all general surfaces, window sills and cupboard tops, with 70% alcohol.
2. In addition to normal washing during use, wash under the work surface of the laminar flow hood (if applicable) with detergent, followed by 70% alcohol.
3. Replace completely the humidity water in the incubator and add a non-volatile germicide, e.g. Roccal II (Jencons Scientific Ltd) or 1% Cetrimide (Cetavlon, ICI).
4. Pour disinfectant down sink outlets.

Every three months:

1. Dismantle the inside of the incubator and wash with detergent and alcohol.
2. Wash prefilter in the laminar flow hood with detergent, rinse, dry and replace.

Also:
Wash gas cylinders before placing in the laboratory.

PROTOCOL 2.3.2: PERSONAL HYGIENE

1. Always wear a clean lab coat, that is reserved for tissue culture. General lab coats should not be taken into the tissue culture laboratory.
2. Outside footwear should not be worn in the laboratory. Preferably keep a pair of shoes especially for use in the laboratory.
3. Unless very short, hair should be covered by a paper hairnet.
4. Try not to sneeze or even talk too much when seated at the laminar flow hood, in order to minimise disturbance of the air currents. If you have a cold, try and avoid tissue culture work, but if this is impossible, wear a paper mask over nose and mouth.
5. Wash hands thoroughly before and after starting work. If you make your own bread or beer, take extra care to clean under fingernails where yeast particles may be lodged. The ICI products Hibiscrub and Hibisol (for washing and final rinse respectively) are recommended. Gloves are not normally recommended since they can reduce dexterity and are uncomfortable if worn for long. They can also give a false sense of security, since they will lose their original sterility as soon as they touch a non-sterile surface.

PROTOCOL 2.3.3: GENERAL TISSUE CULTURE LABORATORY PRACTICE

1. All equipment and supplies not directly used in tissue culture should be kept out of the tissue culture laboratory to facilitate cleaning.
2. Follow the cleaning procedures given above.
3. Do not bring contaminated materials into the tissue culture room.
4. Do not allow any dirty material (used plastic/glassware, waste paper, washes etc.) to accumulate in the laboratory.
5. Restrict access to the tissue culture laboratory to those who are working in it.

Before starting work:

1. Swab work area with 70% alcohol, before and after use, and also between procedures using different cells. Spillages should be wiped up immediately with alcohol.
2. Prepare everything needed for any culture procedure prior to beginning your experiments. Try to minimise interruption of the laminar flow currents by frequent movement of hands in and out of the hood.
3. Arrange only what is necessary for a particular procedure in the hood such that the air flow between the filter and culture is not blocked, i.e. place bottles to the sides, not to the rear, in a horizontal flow hood.
4. Swab media and solution bottles to be used with 70% alcohol before putting them into the hood.
5. Check that every new bag of disposable plasticware is not torn, or has been resealed properly if used previously. Check that bottle tops are tight before opening the bag or before removing from the laminar flow hood.

continued on next page

continued

6. It is advisable to use only your own stock solutions and glassware, not only because you are assured that they were prepared properly, but also to minimise the possibility of cross-contamination.
7. Pour all waste solutions into a disposable container, or one that can be autoclaved. Do not pour waste media down the sink in the tissue culture laboratory before decontamination.
8. Do not work on more than one cell line at the same time and keep media preparation and cell manipulations separate. If a culture is suspected of being infected, work on it after other cell work is finished.
9. Ensure that you are comfortably seated at the hood. Discomfort can lead to sloppy manipulations.
10. Ensure that you have enough time to complete a procedure without undue haste.

After finishing work:

1. Empty hood and wipe surfaces with 70% alcohol.
2. Take all used non-disposable articles out of the tissue culture area to be washed and autoclaved. Empty all waste buckets. Used syringes should be destroyed (i.e. autoclaved) before disposal in general rubbish collections.
3. If packages containing multiple units were started but not finished, seal with tape, label with your name and date, and store in a sterile cabinet.
4. Leave nothing in the culture hood when finished.

PROTOCOL 2.3.4: WASHING AND STERILISING

Glassware:

1. Soak glassware overnight in non-toxic detergent (i.e. 7X, Flow Laboratories Ltd).
2. Wash by machine or bottle brush.
3. Rinse thoroughly in several changes of tap water and distilled water.
4. Dry and cap with aluminium foil before storage.

N.B. The task of washing glassware will be minimised if care is taken not to allow medium to dry inside. Soak in detergent immediately after use.

Sterilise by either:

1. Dry heat (one hour at 160°C). Use within 48 hours.
2. Autoclave (20 min at 120°C). Bottle tops should be loosened a complete turn to allow steam to penetrate, but take care to tighten them after the procedure, as soon as the temperature allows.

N.B. Use indicator tapes on the foil caps, to be sure of the effectiveness of the sterilisation and to avoid mix-ups with unsterilised items.

continued on next page

continued

Dissecting instruments:

1. Wash immediately after use in detergent. Do not soak.
2. Rinse thoroughly in distilled water and dry.
3. Wrap individually in aluminium foil. Label and apply indicator tape.
4. Sterilise by dry heat.

Reusable filter units:

1. Dismantle completely, immediately after use, and soak in detergent.
2. Rinse thoroughly in tap water and distilled water and dry.
3. Place filter paper on base of holder (take care not to scratch/puncture filter paper). Place rubber ring seal centrally on filter paper and carefully attach screw cap (slightly loose). It is important that this does not dislodge the filter paper.
4. Wrap in aluminium foil.
5. Autoclave.

N.B. Tighten before use.

2.4 CULTURE MEDIA

Perusal of the catalogues of companies specialising in supplying tissue culture materials will reveal a daunting range of different growth media. Each of these has been developed to provide optimal growth conditions for particular cell types. Each type of medium is available in liquid form, of single or 10× strength or lyophilised. They are composed of a defined, balanced salt solution with essential amino acids and vitamins and a pH buffering system based on bicarbonate or HEPES. Some will require additional additives before being able to support cell growth but virtually all of them require the addition of animal serum.

The choice of liquid or powder is dependent on several factors. Powders are cheaper, take up less storage space and have a longer shelf life than liquids. However, large volumes of sterile water or filtration equipment will be needed to reconstitute powder media and they will take longer to prepare. The volume of medium kept in store will obviously depend on the volume of work. When ordering medium, bear in mind that liquid culture medium should be kept no longer than one year according to most manufacturers' recommendations, whereas powders can be kept for two years. Medium containing glutamine and serum will only last four weeks, so make up only the volume that you expect to use within that time.

The most commonly used formulation of tissue culture medium suitable for the maintenance of myeloma and hybridoma cells is RPMI 1640. An acceptable alternative medium, often used, is Dulbeccos Minimum Essential Medium (DMEM). Use whichever medium is recommended by the supplier of the myeloma cells and do not change to the alternative medium without a gradual period of equilibration. These media contain all the necessary salts and amino acids for supporting cell growth but require the addition of foetal calf serum before use. They are available commercially in powder form, 10× liquid concentrate or single strength liquid, with or without buffer (bicarbonate or HEPES), with or without glutamine, and with phenol red as a pH indicator. Glutamine is normally only added to powder stock because of its instability in liquid form, but, for cell growth, glutamine must be added before use, and replenished if the medium is kept for more than a month.

PROTOCOL 2.4: PREPARATION OF CULTURE MEDIA: BASIC WASH MEDIUM (one litre of RPMI 1640)

For all temporary manipulations with cells not involving culture, i.e. centrifuging, washing, diluting etc., and as a base for complete culture medium. Single strength medium can be used directly for this purpose. For reconstitution of the concentrated forms, follow the manufacturers' instructions. Where possible work in an aseptic environment (laminar flow hood) with sterile bottles and solutions, even though the finished medium is to be filtered sterile.

Materials:

    Sterile bottle (one litre)
    Sterile water (up to one litre)
    RPMI 1640 powder sachet (one litre) or 100 ml of 10× stock
    $NaHCO_3$ solution (27 ml of 7.5% solution per litre of medium, or 2 g/l)
    0.2-μm filter (see section 2.2.8)

Method:

1. To approximately 800 ml of sterile water in one-litre bottle, add all the contents of the powder sachet or 100 ml of 10× stock and mix thoroughly.
2. Add $NaHCO_3$ solution or powder.
3. Make up to final volume of one litre.
4. Adjust the pH if necessary, with 1 M NaOH or 1 M HCl to 7.4.
5. Immediately filter through 0.2-μm filter into sterile containers with screw cap, leaving minimum air space.
6. Label with name, date, user and store at 4°C.

Comments:

Before general use, test the medium for sterility by keeping a small sample of medium in a sealed bottle at room temperature for a few days. If the medium is to be used as a base for culture (for recipe see Appendix A), use it to maintain some myelomas in culture for a few days. If there is any sign of contamination, autoclave all media and discard.

Avoid multiple sampling of stock solution to minimise risk of contamination, i.e. use 100 ml of 10× stock to prepare one litre of single strength at one time. If all preparation is carried out aseptically with sterile solutions added, there should be no need to filter-sterilise at the end.

For general culture of hybridoma and myeloma cells, three media compositions based on RPMI 1640 (or DMEM) will be required. Firstly, a wash medium or balanced salt solution will be needed for all temporary manipulations with cells not involving culture, i.e. centrifuging, washing, diluting etc. This is simply a buffered, single strength RPMI which is the base for complete tissue culture medium. For cell growth, additives such as glutamine, antibiotics and foetal calf serum (FCS) should be included. FCS is normally used at two strengths, 10% for growth of myelomas and established hybridomas and 20% for fusion and cloning. In order to reduce the considerable expense of FCS, these percentages may be reduced for particular batches of serum but not before adequate trials.

A guide to the preparation of a basic wash medium is described in Protocol 2.4, followed by descriptions of essential and non-essential media additives. Detailed recipes for these can be found in Appendix A.

2.4.1 Essential additives

2.4.1.1   BICARBONATE/CO_2 OR HEPES

The pH of a culture medium is usually maintained by balancing dissolved bicarbonate and CO_2 concentrations. Dissolved CO_2 is regulated at a given temperature, by atmospheric CO_2 tension. If the atmospheric CO_2 is high, there will be more HCO_3 in the medium which readily reassociates with cations, leaving the medium relatively acid. This effect is neutralised by having sufficient bicarbonate already in the medium to maintain the pH at 7.4. Each single strength culture medium is formulated with the appropriate concentration of bicarbonate to buffer adequately at the correct CO_2 tension for a particular cell type. For hybridoma work, RPMI medium has a bicarbonate concentration of 2 g/l to be used at a CO_2 tension of 5%. Concentrated stock media are often kept acid and will need the addition of a specified amount of bicarbonate before use.

Alternatively, media can be buffered with 20 mM HEPES, although for many cell lines there is still a need for atmospheric CO_2 in order to prevent the eventual elimination of dissolved CO_2 and HCO_3^-. RPMI 1640 formulations are available containing HEPES and bicarbonate (Dutch Modification) which are said to be more suitable for lymphocyte stimulation work than ordinary RPMI medium. However, HEPES-buffered media are best avoided at fusion since cell membranes are more permeable to PEG in the presence of HEPES. The addition of sodium pyruvate to the medium enables cells to increase their endogenous production of CO_2, thus reducing the need for exogenous CO_2 and bicarbonate.

These critical buffering requirements will vary according to cell density. Low cell concentrations in open vessels will need to be maintained in a CO_2 atmosphere, whereas for high cell concentrations, especially those in sealed flasks, this may not be necessary.

Phenol red is normally incorporated into culture media as a pH indicator. It is purple at pH 7.8, bluish-red at pH 7.6, red at pH 7.4, orange at pH 7.0, and yellow at pH 6.5.

2.4.1.2 GLUTAMINE

Glutamine, at a working concentration of 2 mM, should be added to all liquid media since it is unstable in solution, having a half-life of only three weeks at 4°C and one week at 37°C. It is normally purchased from tissue culture suppliers as a 100× frozen liquid or lyophilised powder.

2.4.1.3 SERUM

Foetal bovine serum at concentrations usually between 10% and 20% is also an essential additive for hybridoma culture. Lower concentrations are possible with specially formulated 'serum-free' media but it is rarely possible to remove serum altogether during the developmental procedures. However, new formulations of 'serum-free' media are being developed continually (see later), stimulated by the increasing expense of foetal serum.

It is essential to use FBS that has been proved to support clonal growth. Only 10% of general batches have this property. A number of suppliers test serum for clonal growth support, but it is good policy to order a number of samples to compare in your own test (see Chapter 6). Batches of five litres upwards can then be reserved until the test is completed.

2.4.2 Non-essential additives

2.4.2.1 ANTIBIOTICS

For normal culture purposes, a mixed formulation of penicillin (100 U/ml) and streptomycin (100 µg/ml) offers sufficient protection against many Gram-positive and Gram-negative bacterial infections. Regular use of other antibiotics (see section 2.5) is not recommended, to guard against the possibility of encouraging resistant strains of micro-organism. Indeed, it has been argued that even 'Pen and Strep' should be omitted, so that the practice of high standards of aseptic technique is maintained.

2.4.2.2 2-MERCAPTOETHANOL

A 50 µm concentration of 2-ME added to the culture medium enhances antibody synthesis by spleen cells and renders hybridoma cells less dependent on particular batches of serum (Click et al, 1972). The effect is probably mediated by the lipid antioxidant activity of reduced glutathione, the availability of which is enhanced by 2-ME (Hoffeld & Oppenheim, 1980; Hoffeld, 1981).

2.4.2.3 GROWTH FACTORS

More and more specific growth factors for particular cell lines are being introduced. Those available for hybridoma stimulation will be discussed in Chapter 6.

2.5 INFECTION: RECOGNITION, PREVENTION AND ELIMINATION

The practice of proper aseptic techniques, as described above, and the inclusion of appropriate antibiotics in the culture medium, significantly reduce the likelihood of infection of cultures with micro-organisms. Nevertheless, there are so many potential sources of contamination that infection is bound to occur at some time. Being able to recognise these infections and to deal with them rapidly before they have time to spread is, therefore, of paramount importance. As a general rule, any cell culture found to be contaminated should be discarded immediately, including suspect media or additives. On occasion, when valuable long-term cultures become contaminated, it may be worth trying to eliminate the infection. This is possible in certain circumstances but is never easy. The various types of contamination commonly encountered in tissue culture laboratories and ways of dealing with them are described below. If none of these measures is effective and the contamination persists after all precautions are taken, it is probably worth getting the infection positively identified by a microbiology laboratory, together with swabs of areas to which the cells have been exposed. Specific antibiotics can then be employed and hopefully the source of the contamination isolated.

2.5.1 Bacteria

Bacterial contamination is probably the easiest to recognise. The medium becomes acid (yellow), and opaque, and on microscopic examination the cells are obliterated and covered in a sea of rapidly moving tiny black specks (see Fig. 2.5.1). Growth of bacteria is very rapid, increasing from apparently nothing to overwhelming the culture in less than 24 hours. If the cells are unimportant or if there are frozen stocks, discard the culture and start again. Many culture media routinely contain penicillin (100 U/ml) and streptomycin (100 μg/ml) which are effective against Gram-positive and Gram-negative

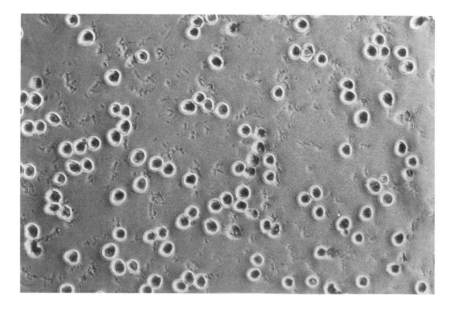

Fig. 2.5.1 Bacterial infection of myeloma cells. Magnification ×944

bacteria respectively. Infection arises in such media due to the development of resistant strains. A third antibiotic, gentamicin (200 μg/ml), is effective against a broader spectrum of both Gram-positive and Gram-negative bacteria, including *Pseudomonas* strains, and can overcome some strains of *Proteus* and *Staphylococcus* that can develop resistance to 'Pen & Strep'. It is inadvisable to culture indefinitely in gentamicin, since this would encourage the evolution of gentamicin-resistant strains. If a few individual wells in a multiwell plate become contaminated, and the remainder are worth saving, carefully aspirate the contaminated wells and replace with 3 M NaOH. If the facilities are available, it is also possible to decontaminate a bacteria-infected culture using the fluorescence activated cell sorter.

2.5.2 Fungus and yeast

Fungal contamination varies according to laboratory environment and is more common in old buildings with inefficient air-conditioning systems. It also varies with season, being most common in late spring and early summer in this area. It is rather difficult to visualise at first, especially when the cells appear viable (see Fig. 2.5.2a). Eventually a fluffy ball appears on top of the culture. If this infectious stage is reached, discard the culture immediately. Further contamination can be discouraged by the use of amphotericin B in the medium. The concentration is critical; 2.5 μg/ml of medium is recommended but doses as low as 10 μg/ml can be toxic to certain cell types. Dose control is further complicated by the short half-life (five days) of amphotericin.

Yeast contamination can also be avoided by the use of amphotericin B or nystatin (50 U/ml). It is easily recognised as small 3–5-μm diameter oval or spherical cells often arranged in short chains (see Fig. 2.5.2b).

2.5.3 Mycoplasma

Mycoplasmas also known as PPLOs (pleuralpneumonia-like organisms), are probably the most serious type of contamination for long-term cultures, since they are very difficult to detect microscopically and have a range of effects on cell growth from none, to retardation or cell death. They compete with the cell for nutrients, particularly nucleic acid precursors, and so can have profound effects on macromolecular synthesis (including radioisotope incorporation studies), stability of genetic material and the response of cells to drugs or selective media such as HAT medium. There may also be difficulties in recovering cells from frozen storage. Mycoplasma infections are far more common than most people realise. Some estimate that as many as one in six continuous cell lines are contaminated. It cannot be emphasised too strongly that cultures should be regularly tested for the presence of mycoplasma and so should all new cultures entering the laboratory. In addition, you will not be thanked for sending infected cultures to other laboratories. It should also be noted that regulatory bodies such as the US Food and Drug Administration advise that all cell culture products used for diagnostic kits or therapeutic reagents derived from cultures are free from mycoplasma.

There are at least 30 named species of these small (130–300 nm) prokaryotes without a cell wall which attach to the membranes and grow in the cytoplasm of mammalian cells. They are very pleomorphic, ranging from coccoid to filamentous, and can form

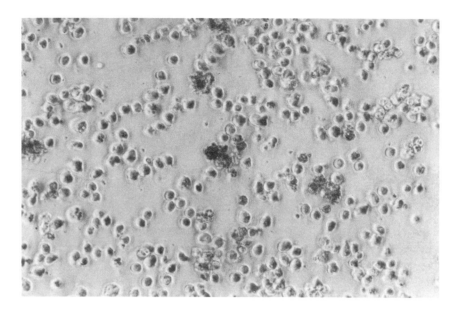

Fig. 2.5.2a Fungus infection of myeloma cells. Magnification ×944

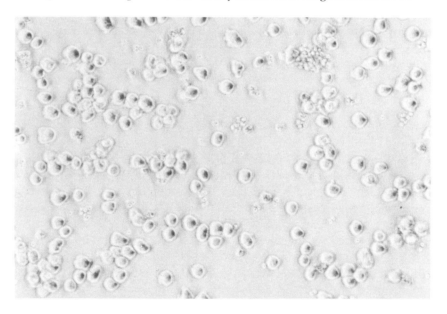

Fig. 2.5.2b Yeast infection of myeloma cells. Magnification ×944

small colonies from 100 μm to 1 mm in diameter. The majority of mycoplasma infections in cell culture are caused by five species: *Acholeplasma laidlawii* (bovine), *Mycoplasma fermentans* (human), *Mycoplasma hyorhinis* (porcine), *Mycoplasma orale* (human), and *Mycoplasma pirum* (natural habitat unknown). The likely sources of contamination are inefficient aseptic technique, serum, or an infected cell line. Simple preventative measures should be employed routinely, such as separate handling of different cell lines, including disinfection of hood surfaces and a delay of 10 minutes and handwashing between handling.

There are many methods by which contamination can be detected, either biochemically, microbiologically, by electron microscopy or by fluorescence methods. Each has its own limitations so it is advisable to try more than one test. One of the most widely used employs a fluorescent DNA stain, Hoechst 33258 (Calbiochem), also available in kit form from Flow Laboratories (Chen, 1977). A protocol for mycoplasma testing with Hoechst stain is given below (Protocol 2.5.3). A positively infected culture is needed to control the test since interpretation of fluorescent slides requires skill. Care must, therefore, be taken not to introduce new infection into the laboratory.

PROTOCOL 2.5.3: TEST FOR MYCOPLASMA INFECTION WITH HOECHST STAIN

Materials:

Glass coverslips
35-mm culture dishes
Test cell sample
5% CO_2 incubator
Pasteur pipettes
Carnoys fixative (3 : 1, methanol : glacial acetic acid)
PBS for washing and stain dilution
Hoechst stain (prepare 1 mg/ml diluted to 1 : 1000 as stock)
Mountant (Moviol 4-88, see Appendix A)
Glass slides
UV epifluorescent microscope with appropriate filters

Method:

1. Sterilise coverslips (i.e. in 70% ethanol), air dry and place in 35-mm culture dishes.
2. Seed test cell sample into dishes at approximately 5×10^5 cells/ml in culture medium without antibiotics. Set up two cultures for each sample.
3. Incubate in a 5% CO_2 incubator at 37°C for 24 hours.
4. Remove one coverslip from each pair of samples. Fix in Carnoys fixative (i.e. 5 min) and air dry. Take care to apply fixative round the rim of the dish, to minimise physical detachment of cells. Leave the other coverslip to incubate for a further 48 hours before fixing.
5. Stain each coverslip with Hoechst stain at a final dilution of 0.05 μg/ml for 5 min in the dark.
6. Wash in PBS and mount on glass slides with Moviol mountant.
7. Examine by UV epifluorescence at 100× magnification. Clean cultures will show only nuclear staining but infected cultures will also show staining of mycoplasma in the cytoplasm and extracellularly (Fig. 2.5.3).

Comment:

The filter combination used on the fluorescent microscope is important. We use an excitation filter of 340–380 nm, and a suppression filter of 430 nm with a light filter block A.

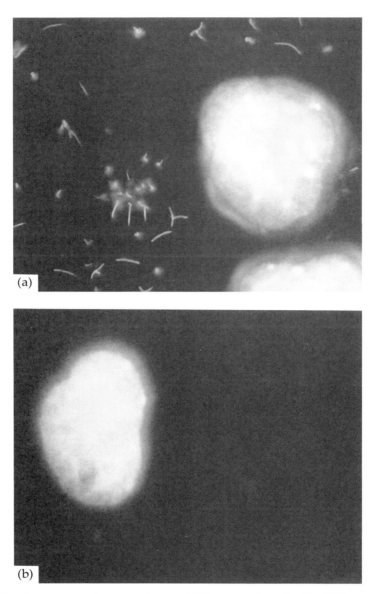

Fig. 2.5.3 Hoechst stain test for mycoplasma. (a) Uncontaminated cells. (b) Extracellular staining of mycoplasma-contaminated cells. (Photograph courtesy of Jon Mowles, ECACC)

An alternative test (GIBCO), that does not need viable mycoplasma as a control, is based on the high level of adenosine phosphorylase in mycoplasma relative to mammalian cells (Hatanaka et al, 1975). Adenosine phosphorylase converts 6-methylpurine deoxyriboside (6-MPDR), a non-toxic analogue of adenosine, into 6-methylpurine and 6-methylpurine riboside, both of which are toxic to mammalian cells. In the presence of this substrate infected cells will die, because of the presence of the enzyme. Clean cells should survive since there is no enzyme present. A positive control is provided by the addition of purified enzyme. The disadvantages of this test are that some mycoplasmas may not be detected, antibiotics may give false negatives and certain organisms with higher levels of adenosine phosphorylase will give false positives.

A third test is based on the incorporation of tritiated thymidine into cells. Mycoplasmas deplete exogenous thymidine (or uridine) by enzymatic degradation into free base and sugar moieties. They will, therefore, compete with cells for thymidine, resulting in inhibition of radiolabel entry into cells. This effect can be measured in one of three ways: by harvesting the cells and counting label incorporation; by autoradiography, showing a similar staining pattern as described with the fluorescent stain; or by measuring chromatographically the cleavage of $[^3H]$ thymidine to $[^3H]$ thymine by infected medium (Perez et al, 1972).

A new test kit has recently been introduced by Laboratory Impex Ltd, 'Gen-Probe Mycoplasma T.C. II Rapid Detection System', based on the use of a tritium-labelled DNA probe homologous to *Mycoplasma* and *Acholeplasma* ribosomal DNA. The test takes only two hours to perform, and claims to detect all species that commonly infect tissue cultures. Non-infectious nucleic acids are included as positive controls.

Eradication of mycoplasma infection is extremely difficult and is not worth the effort or risk to other cells unless the infected cells are particularly valuable. The antibiotics reported as being of particular use are tanamycin, tetracyclines, tylosin, erythromycin, and lincomycin, although resistance can easily be induced. Combinations of these with activated macrophages has also had effect and so too has elevation of incubation temperature to 41°C. In the specific area of hybridoma production, passage of contaminated cultures through mouse ascites followed by subcloning has been effective. Recent reports suggest that the use of ciprofloxacin, one of a new group of antibiotics called fluoroquinolones, may prove to be the most effective strategy (Schmitt et al, 1988; Mowles, 1988).

A mycoplasma testing and identification service is offered by the European Cell Culture Collection or the American Type Culture Collection (Appendix C). An elimination service is also offered at considerable cost which may be justified if the cell line is particularly valuable.

Chapter 3

Immunisation

3.1   PRINCIPLES

A major prerequisite for the production of monoclonal antibodies is a supply of antibody-producing cells, from among which those secreting specific antibody can be selected. The purpose of this chapter is to outline the practical considerations relevant to acquiring lymphoid cells that are not only sensitised to the antigen in question but are also available in such numbers and in such a state that they are useful for monoclonal antibody production. For this purpose, two approaches are possible. The first involves exposing cells to the foreign substance while they are in their normal environment, i.e. in vivo. The second involves the removal of presensitised or unsensitised cells from the donor with subsequent exposure to antigen in vitro. The various procedures required to sensitise cells are the substance of the following sections and will deal with the animal donors and the ways in which materials of differing immunogenicities may be administered in order to produce a useful immune response. The less well established in vitro procedures, which are enjoying a growing popularity, will be described in the latter part of the chapter.

3.2   THE LAW

In the UK, the administration of any foreign material to an animal is a procedure requiring registration both of the operator and the premises in which the work is to be carried out. The production of both polyclonal and monoclonal antibodies is covered by the requirements of the Animals (Scientific Procedures) Act 1986. The introduction of this new Act of Parliament in 1986 has drastically altered the legal framework under which such work is now carried out.

Thus, the level of protection afforded to experimental animals used in the UK has been extended to include provision for a named person in the day-to-day care of such animals, together with regular inspection of the health and welfare of both experimental and stock animals by a practising veterinary surgeon. These new safeguards are in addition to the existing checks carried out by the Home Office Inspectorate.

In addition, the method of granting licences to individuals by the Home Office to perform these procedures has been altered. Thus, where previously a personal licence was issued as the sole permission to use such animals, the new Act now requires a project licence which sets out in detail the animal species (including some idea of total numbers required), the techniques involved, and an appreciation of the potential pain levels that may possibly be felt by the animals concerned. In addition, the project licence must include the names of all those who will perform such procedures under this licence.

Therefore, a project licence, covering all these aspects of work, as it pertains to the production of antibodies and specifically the production of polyclonal sera and monoclonal antibodies, will be required together with the possession of personal licences covering the individuals who will be involved. It is essential that all who intend to use animals as part of research projects within the UK clearly understand their own inherent responsibilities under this Act, and those in doubt are advised to contact the Home Office (see Appendix C).

In the US, the basic legislation covering the use of animals in scientific research is embodied in the Laboratory Animal Welfare Act which, since its enactment in 1966, has been the subject of substantive amendments in 1970, 1976 and 1985. Importantly, the 1985 amendments emphasised the need to control the extent of pain and suffering and are supplemented by NIH guidelines which require those receiving funds to comply with the provisions of the Public Health Service Policy on Humane Care and Use of Laboratory Animals (US Department of Health and Human Service, 1985). In the operation and application of these guidelines, institutions are required to establish local regulatory procedures, and although many of the guidelines do not cover rats and mice those intending to produce antibodies should consult with their local institutional Animal Welfare Committee before commencing work (see Rollin, 1987). This latter publication also summarises the legislative constraints governing the use of animals in biomedical research for the 21 member countries of the Council of Europe.

Although the maintenance of complete records of animal usage are a requirement of the law in some countries, it is essential, if only for the purposes of good laboratory practice, to always keep such records and to ensure that all relevant information is kept and stored safely.

All experimenters should comply with the requirements of the humane use of animals regardless of the legislation that may surround that use and all new workers, who may not have used animals previously, should consult with more experienced colleagues and read the standard texts on animal welfare and handling, e.g. Tuffery (1987), Herbert et al (1986), Universities Federation for Animal Welfare (UFAW) Handbook, Kelly et al (1988), and Institute of Animal Technology (IAT) Videos.

3.3 ANIMALS

In practice, the majority of readers will be using either mice or rats as the source of sensitised lymphocytes, with the former species being pre-eminent. In addition, larger species, such as rabbits, may be required as antiserum donors, to provide positive controls in screening tests and characterisation assays.

The choice of rat or mouse strains for use in monoclonal antibody production is normally constrained by the sources from which the myeloma cell lines were derived. Thus, the Balb/C mouse strain is preferred since many of the tumours from which myeloma cell lines have been derived were originally induced in such animals and they are therefore compatible for the propagation of hybridoma cells in vivo. Similarly, the Lou strain of rats is the preferred immune lymphocyte donor.

However, this constraint on the choice of strain need not be absolute: firstly, ascites production in vivo is not the only procedure for the propagation of hybridomas; and secondly, if alternative strains have to be used as lymphocyte donors, ascites production is possible if F1 hybrid animals crossed with Balb/c are used (see Chapter 6).

The animals required for this work may be purchased as required from registered suppliers (see Appendixes B and C). However, because immunisation protocols take several weeks and animals of various ages will be required, maintenance facilities will be needed. Therefore, a breeding colony dedicated to the needs of monoclonal antibody production would be both convenient and cost effective.

Such a colony will need to satisfy the demands for ongoing immunisation programmes, the production of ascitic fluid and animals as donors of feeder cells and non-immune serum. There is no restriction on the sex of the animals to be used for these procedures, although female animals are preferred for immunisation since they have a more placid nature when housed in groups for extended periods. Pragmatically, therefore, it is often useful, in order to keep the colony balanced, to use male animals for ascites production. Animals of between 8 and 12 weeks of age are normally required for the initiation of a particular immunisation protocol, whereas full-grown adults, aged three to six months, are normally used for ascites production. If thymocytes are to be used as feeder layers (see Chapter 6) it is essential to have an assured supply of younger animals (up to 10 weeks) as thymus donors.

If it is planned to raise interspecies or human hybridomas and to propagate antibody in vivo, it may be necessary to maintain a colony of nude mice. These are hairless athymic mouse mutants, first described in detail by Flanagan in 1966. They have been useful natural athymic models in the study of thymus function and models for the in vivo growth of neoplasms. Their immuno-incompetence, therefore, provides an ideal environment for the growth of interspecies hybridomas. However, one drawback to their use is the need for special, aseptic conditions in which to house them. Nude rats have also been described (Festing et al, 1978).

3.4   ANTIGENS AND IMMUNOGENS

An antigen is the molecular entity which has the capacity to react with or bind selectively to an immunoglobulin. Presumably, even at an early stage of any new work in this area, the overall nature of the antigen will be known, although in some cases this knowledge may be of a very general and operationally based nature; e.g. putative cell surface (differentiation) antigens. For the production of monoclonal antibodies, and in common with the production of polyclonal antisera, the antigen is often also the immunogen. The immunogen is the material which has the capacity, upon administration to the recipient, to elicit antibody production. Some substances can be antigens without being immunogenic and in general only macromolecules have the two properties of antigenicity and immunogenicity. For a substance to be immunogenic it must be a relatively complex structure with a composition which is recognised as foreign by the recipient. The classes of molecules and structures most commonly used as immunogens cover an enormous range of natural and synthetic materials including proteins, nucleic acids, carbohydrates, lipids, cellular components, bacteria, viruses and a wide variety of xenobiotics. Most of these are composed of numerous antigenic determinants (epitopes), that is, sequences of continuous or folded molecular structure, each of which may be capable of interacting with a specific antibody. Whether these epitopes are capable of eliciting an antibody response will depend on their recognition by the host as foreign and also on their molecular properties.

A number of factors have been investigated, particularly for proteins, with regard to the relationship between molecular properties and immunogenicity. Molecular size plays an important role. Generally, low molecular weight antigens (less than 1000) are not good immunogens. With proteins, both the amino acid composition of the immunogenic determinant and its accessibility will play a part. The overall conformation of the molecule is important, particularly the disposition of active groups at the surface of the macromolecule rather than in the interior of the structure.

When raising polyclonal antisera to large complex molecules, the antibody response to dominant epitopes is likely to mask the response to other epitopes which might be of more interest. Although there are immunisation strategies to overcome this problem, in monoclonal antibody production efforts to isolate antibodies against minor epitopes are usually concentrated on refining the specificity of the post-fusion screening tests, not on pre-fusion immunisation.

3.4.1 Crosslinking, conjugation and haptenisation

Greater difficulties are often encountered at the other end of the scale when trying to raise antibodies against substances that may not be naturally immunogenic. Various chemical modifications are possible depending on molecular structure (for review see Erlanger, 1980). These include crosslinking monomers into supramolecular complexes or conjugation of the molecule to carrier proteins or haptens. Haptens are small molecules which can bind to B-cells but cannot stimulate the B-cell to differentiate and produce antibody. The co-operation of T-cells is usually required for this and haptens are too small to stimulate both cells simultaneously. However, if the hapten is conjugated to a larger immunogenic molecule, this problem can be overcome and antibody production against both the hapten and carrier molecule is stimulated.

A number of factors require consideration, including the choice of carrier molecule and coupling reagent, the method of coupling and which functional groups to use in the coupling. Common carrier proteins include albumin, immunoglobulin and keyhole limpet haemocyanin. Serum albumins have the advantage of being more soluble than γ-globulin or ovalbumin (insoluble conjugates, however, may be useful when purifying antibody).

Common coupling reagents are glutaraldehyde and carbodiimide (other examples are also given in Table 3.4.1). Ideally, one should aim for between 8 and 25 hapten molecules per carrier molecule to obtain reasonable antibody titres. It has long been recognised that antibody specificity is directed primarily at the part of the hapten molecule furthest removed from the site coupled to carrier. The introduction of a spacer arm to separate hapten from carrier can also improve antibody specificity.

The groups on carrier molecules that have been used to attach to haptens to produce immunogenic conjugates include amino groups of the N-terminal and lysine residues, carboxyl groups of the C-terminal and aspartic and glutamic residues, imidazo and phenolic functions of histidine and tyrosine residues respectively, and sulphhydryl groups of cysteine residues.

Table 3.4.1 Coupling reagents

Name	General structure	Groups modified	Linking	Bonds formed
Carbodiimide	$R_1-N=C=N-R$	$-COOH$	$-NH_2$	$\begin{array}{c} O \\ \parallel \\ -C-N- \\ \mid \\ H \end{array}$
Glutaraldehyde	$HCO(CH_2)_3CHO$	$-NH_2$	$-NH_2$	Schiff base
Mixed anhydride	$R_1.CO_2COR_2$	$-NH_2$	$-NH_2$	$\begin{array}{c} O \\ \parallel \\ -C-N- \\ \mid \\ H \end{array}$
Bifunctional imidates	$(CH_3OCN^+H_2(CH_2)_nCN^+H_2OCH_3)2Cl^-$ $-NH_2$	$-NH_2$	$\begin{array}{c} {}^+NH_2Cl^- \\ \mid \\ -C-N- \\ \mid \\ H \end{array}$
Bifunctional succinimide/ maleimide	$\begin{array}{c} O \quad\quad O \quad\quad\quad O \\ \parallel \quad\quad \parallel \quad\quad\quad \parallel \\ \rangle NOC-(R)-N\langle \\ \parallel \quad\quad\quad\quad \parallel \\ O \quad\quad\quad\quad\quad O \end{array}$	$-NH_2/-SH$	$-NH_2/-SH$	$\begin{array}{c} O \\ \parallel \\ -C-N- \\ \mid \\ H \end{array}$

Protocols of the commonest conjugation methods are given below (Protocols 3.4.1 and 3.4.2). It should always be borne in mind, however, that such chemical modifications may irretrievably alter a particular epitope that a monoclonal antibody might recognise.

On testing the final antibody product, use free hapten or a different conjugate with different carrier and coupling reagent, to eliminate the possibility of cross-reactivity with anything other than the specific antigen.

PROTOCOL 3.4.1: CONJUGATION OF KLH OR BSA TO A PEPTIDE (ONE-STEP GLUTARALDEHYDE METHOD) (after Avrameas & Ternynck, 1969)

Materials:

Peptide
KLH, 1 mg/ml (1 M KLH = 3×10^6 mg/ml)
  or BSA, 1 mg/ml (1 M BSA = 67×10^3 mg/ml)
PBS, pH 7.35, for dilution and dialysis
Glutaraldehyde (i.e. grade I, 25% aqueous solution, Sigma)
L-Lysine, 1 M
Glycerol for storage

continued on next page

continued

Method:

1. Add peptide to a 1 mg/ml suspension of KLH or BSA at a molar ratio of carrier protein to peptide of 1 : 20 to 1 : 40. Dilute solutions in PBS.
2. Add an equal volume of glutaraldehyde (0.1–2% v/v) dropwise with constant stirring.
3. Incubate at room temperature for two to three hours.
4. Stop the reaction by addition of 1 M L-lysine at 1 : 20 volume.
5. Dialyse extensively against PBS.
6. Store at −20°C after adding an equal volume of glycerol.

Comment:

KLH is not very soluble in aqueous solutions. Dissolve as much as possible and measure protein present (OD at 280 nm). Adjust other concentrations accordingly. This method of conjugation is poorly controlled and can result in the formation of large polymers. For immunisation, however, this is not a problem.

PROTOCOL 3.4.2: CONJUGATION OF CARRIER TO PEPTIDE (CARBODIIMIDE METHOD) (after Goodfriend et al, 1964)

Molar ratio of carrier to peptide 1 : 100.

Materials:

Peptide
Carrier (i.e. KLH or BSA as above)
Carbodiimide (CDI; 1-ethyl-3-(3-dimethylaminopropyl) carbodiimide)
Dilute HCl or NaOH for pH adjustment

Method:

1. Mix 1 mg of carrier and 100 µg peptide in distilled water.
2. With constant stirring, add 10 mg CDI in three aliquots, keeping pH between 4 and 6 with dilute HCl or NaOH.
3. Leave in fridge at 4°C overnight.
4. Incubate at room temperature for two hours.
5. Dialyse against water.
6. Freeze-dry if necessary to reduce volume for injection.

Comment:

These conjugates are unstable, lasting only a few days in solution.

3.5 ADJUVANTS

Despite any indicators of immunogenicity that may be gleaned from precedent or theory, in practice most immunogens are administered together with substances known to augment the recipient's response. Such substances are known as adjuvants and many different materials have been claimed to exhibit the desired action.

Three main classes of adjuvant have been identified. Firstly there is a group known as the 'antigen depot' adjuvants. The most commonly used agents of this type include Freund's complete (CFA) and incomplete (IFA) adjuvant and preparations that contain aluminium hydroxide. Primarily, these adjuvants act by facilitating slow release of the immunogen from the point of administration. The oil-based adjuvants of this class, like Freund's (CFA and IFA) when administered in the form of stable emulsions containing immunogen, also act to induce the formation of a granuloma which is rich in macrophages and immunocompetent cells. Furthermore, because the droplets containing adsorbed immunogen become transported in the lymphatics, these materials also have a direct effect on the delivery of the immunogen to the antibody-producing cells. A protocol for producing a stable emulsion is given below (Protocol 3.5). Complete Freund's adjuvant is distinguished by the presence of a heat-killed preparation of *Mycobacterium tuberculosis* or *M. butyricum* suspended in the light mineral oil base which also contains emulsifying agents. The presence of the *Mycobacterium* enhances the local inflammatory response and for this reason secondary inoculations of immunogen should be given in the presence of incomplete Freund's adjuvant from which the *Mycobacterium* is excluded.

The second major group of active preparations are the bacterial adjuvants. Examples of this type of adjuvant include *Corynebacterium parvum*, *Bordetella pertussis* and the synthetic fragment of bacterial cell wall peptidoglycan, muramyl dipeptide (MDP). These agents probably stimulate macrophages directly. The third group of adjuvants is made up of amphipathic and surface active agents, the most widely used of which is saponin. The efficacy of this adjuvant is particularly noticeable when membrane components are used as the immunogen.

Liposomes can also be used as adjuvants and have particular applications when dangerous immunogens or high doses of immunogen administered intravenously have to be dealt with.

PROTOCOL 3.5: PRODUCTION OF A STABLE EMULSION OF IMMUNOGEN IN FREUND'S ADJUVANT

It is advisable to wear gloves and eye protection when handling Freund's adjuvant.

Materials:
   Freund's adjuvant (complete for first injection, incomplete for second and
      subsequent injections)
   Immunogen
   Capped conical tube to hold total volume
   Vortex mixer
   Beaker of water for stability test

continued on next page

continued

Method:

1. Calculate the total volume of emulsion to be injected, based on maximum volumes recommended in Table 3.6.1, and number of animals to be injected.
2. Dispense half that volume of adjuvant into a small capped tube.
3. Prepare an equal volume of the aqueous immunogen preparation at the appropriate concentration and add it to the adjuvant, slowly and dropwise, with vigorous mixing between drops (i.e. on a vortex mixer). It is important to use a capped tube in this procedure.
4. The final preparation will be very viscous and will require patience in transferring to a syringe for injection. It will not pass through a 25-g needle. As a result there is likely to be a loss of volume so the starting volumes should be adjusted to compensate.
5. Test for stability as described below.

Alternative method:

1. Place equal volumes of adjuvant and immunogen in separate 1-ml syringes with 21-g needles.
2. Join the syringes by attaching to the needles a 15-cm length of fine bore plastic tubing such that the tubing fits tightly to the needles.
3. Hold a syringe in each hand and gently apply pressure to one barrel while releasing the pressure on the other.
4. Repeat this manoeuvre in both directions until an emulsion is formed.
5. Test for stability as described below.

Comment:

This alternative method has the advantage that the emulsion is already in the syringe ready for injection, but there is a high risk of the syringe connections bursting under pressure and spilling the contents, if sufficient care is not taken.

Another acceptable alternative involves sonication, but this method is best used for larger volumes. There is also a possibility that the immunogen may be damaged by sonication.

Testing the stability of the emulsion:

Apply a small drop of the emulsion to the surface of a 100-ml beaker of water. If the drop remains as a discrete droplet and does not spread across the surface, the emulsion is stable enough to inject.

3.6   ROUTES OF ADMINISTRATION

The anatomical site at which immunogens are administered also contributes to the overall antibody response. The aim is to create a slow release depot of immunogen, so the site should not be in highly vascular areas that can quickly remove and metabolise the immunogen. In mice and rats, the most appropriate site is subcutaneous, at the scruff of the neck. Subcutaneous injection is also suitable for rabbits, but because of the larger surface area, multiple injection sites are preferred. Intradermal and intramuscular injections are also common routes in the rabbit, but are not normally used in smaller rodents because of their size and skin thickness. Foot pad injections are said to produce good immune responses since they are close to lymphatic systems, but are not advised because of the discomfort caused to the animal. Prefusion boosts are usually given via the intravenous route in order to quickly stimulate B-lymphocyte proliferation. Recommended inoculation routes, maximum volumes and bleeding methods are given in Tables 3.6.1 and 3.6.2; however, if immunising animals for the first time, it is recommended that these procedures are learnt from watching experienced operators.

Table 3.6.1   Recommended inoculation methods

Route of administration	Maximum volume	Needle size	Restraint
Mice			
s.c.	0.2 ml	21 g	Hand*
i.p.	0.5 ml	21 g	Hand
i.v. (tail)	0.2 ml	25 g short	Ether/box
Rats			
s.c.	0.5 ml	21 g	Hand
i.p.	5 ml	21 g	Hand
i.v. (tail)	0.5 ml	25 g short	Ether
Rabbits			
s.c.	5×0.2 ml	21 g	Hand/box
i.d.	10×0.1 ml	21 g	Hand/box
i.m.	2×0.5 ml	21 g	Hand
i.v. (ear)	1 ml	25 g	Hand/box

*Mice can easily be restrained and injected by one operator.

Rats and rabbits are best held and injected by different operators unless a restraint box is available.

Needle sizes are those recommended for injection of an adjuvant emulsion or cell suspension. Intravenous injections are made with physiological buffers.

Table 3.6.2   Bleeding methods

These volumes represent single bleeds of adult animals with recovery. If repeated bleeds are required take smaller volumes and allow time between bleeds for recovery (preferably weeks). Take care not to cause excessive scarring at bleed sites. If bleeding from the retro-bulbar plexus, use one eye only. Larger blood volumes can be obtained if the animal is to be killed.

Species	Site	Volume
Mice	Retro-bulbar plexus	Up to 0.5 ml*
	Cardiac puncture	Up to 0.7 ml
Rats	Retro-bulbar plexus	Up to 2 ml*
	Cardiac puncture	Up to 5 ml
Rabbits	Marginal ear vein	Up to 30 ml

*It should be noted that some Home Office Inspectors in the UK forbid the use of this method of bleeding.

3.7 IMMUNISATION PROCEDURES

The timing of immunisation protocols is based on our knowledge of the development of an antibody response in vivo. The antibody response to primary injection of an immunogen follows a characteristic pattern. Initially, there is a lag phase, followed by a logarithmic increase in antibody levels, then a plateau before levels decline. The whole response is spread over approximately three weeks. The antibodies produced are predominantly IgM. If a secondary challenge with immunogen is presented three to four weeks after the first, then the antibody response is quicker, much greater and lasts for longer. The predominant class of antibody is IgG and the average affinity is increased, particularly if a low immunogen dose is given.

Table 3.7 shows examples of immunogen doses that have been used in the production of monoclonal antibodies. Protocols tend to follow the simple pattern described above (see Protocols 3.7.1 and 3.7.2 for rabbits and mice respectively) although, where previously uninvestigated immunogens are involved, experiment and chemical modification may be required to produce optimal procedures.

Once a satisfactory antibody response has been established, animals can be kept as a pool of immune lymphocyte donors for fusion. In practice, it has been shown that the newly challenged and thus rapidly dividing cells preferentially fuse to produce hybridomas. Success is best assured if the final booster injection is given three to four days prior to harvesting the lymphoid cells. If the cells are allowed to reach the peak of antibody production in vivo, they are beyond the best state for fusion.

The immunogen preparation need not necessarily be pure for the successful production of monoclonal antibodies. However, to ensure the best chance of yielding specific hybridomas, the presence of the desired antibody should be detectable in the serum of the recipient prior to the final booster dose. Ideally, the specific antibody titre should be at least 1 : 1000 and preferably more.

Bacteria and viruses are invariably attenuated in some way before use as immunogens which may involve treatment of the organism at elevated temperature under anaerobic conditions. Bacterial toxins, such as those produced by *Diphtheria* and tetanus bacilli, may be detoxified by treatment with formaldehyde which does not destroy the major immunogenic determinants.

Table 3.7 Recommended doses of immunogen

Immunogen	Preliminary doses	Final booster dose
Soluble and membrane proteins	10–100 μg	Up to 500 μg
Nucleic acids	200 μg	200 μg
Eukaryotic cells	2–20 $\times 10^6$	2–20 $\times 10^6$
Bacterial cells	50 μg of protein	50 μg of protein
Viruses	10^7 particles (three doses at weekly intervals)	
Fungal antigens	20–100 μg	20–100 μg

PROTOCOL 3.7.1: IMMUNISATION OF RABBITS

Materials:

Immunogen (i.e. 10–50 μg soluble protein per dose, in emulsion with CFA or IFA)
Syringe and 21-g needle

Method:

1. Inject recommended volume at multiple sites (see Table 3.6.1) intradermally, subcutaneously or intramuscularly.
2. Inject secondary dose three to four weeks after primary.
3. Test bleed seven to ten days after secondary dose.
4. Repeat secondary dose and bleeds if antibody titre is unsatisfactory.

PROTOCOL 3.7.2: IMMUNISATION OF MICE

Materials:

Immunogen (Soluble protein: 10–50 μg per dose, in emulsion with CFA for primary dose or IFA for subsequent doses. Cells: 10^6 per dose in PBS)
1-ml syringe and 21-g needle

Method:

1. Inject maximum 0.2 ml of immunogen preparation subcutaneously.
2. Inject secondary dose three to four weeks after primary.
3. Test bleed seven to ten days after secondary dose.
4. Repeat dose and bleeds if antibody titre is unsatisfactory.
5. Give final intravenous booster injection, without adjuvant, three to four days before fusion.

3.8   IN VITRO IMMUNISATION

In vivo immunisation, despite all the enhancement procedures described above, often fails to stimulate adequate numbers of antigen-specific B-lymphocytes. The reasons for this may be induction of tolerance or an overwhelming response to other immunogens or dominant epitopes in the immunogen preparation. In efforts to overcome these problems, the activation of B-lymphocytes in culture, in response to a specific antigen, is receiving increasing attention. The additional advantages of the procedure over conventional in vivo immunisation are that very small amounts of immunogen can be used, as can immunogens of potential toxicity to the whole animal, or self-antigens, and the activation process is considerably shorter. However, potentially even greater advantages are possible in applying the technique to the stimulation of human lymphocytes which can rarely be stimulated in vivo against specific antigens.

There have been numerous reports, in recent years, of success with in vitro immunisation both in the rodent systems and human (for review see Reading, 1982; Borrebaeck, 1988). A variety of different methods and conditioned media additives have been recommended. There is a tendency to produce predominantly IgM antibodies, and any IgG antibodies are of generally lower affinity than those obtained by in vivo methods. A far greater proportion of IgG antibodies can be obtained by stimulating, in vitro, splenocytes that have been primed in vivo.

One of the simplest in vitro methods is described by Boss (1984), the essentials of which are given in Protocol 3.8.

PROTOCOL 3.8: IN VITRO IMMUNISATION (after Boss, 1984)

Materials:

Spleen cell suspension
20 ml of complete culture medium
   (i.e. RPMI 1640 + 20% FCS, 5×10^{-5} M 2-mercaptoethanol, 2 mM glutamine,
   1 mM sodium pyruvate, 100 U/ml penicillin, 100 μg/ml streptomycin)
250-ml tissue culture flask
Immunogen (see Table 3.8 for dose and pretreatment conditions)
N-acetylmuramyl-L-alanyl-D-isoglutamine (MDP), 20 μg/ml final concentration

Method:

1. Prepare a single cell suspension of spleen as described in Chapter 5.
2. Resuspend the spleen cells at a density of 10^7 cells/ml in 20 ml of complete culture medium in a 250-ml tissue culture flask.
3. For sensitisation, add 20 μg/ml of MDP and immunogen as detailed below.
4. The cell/immunogen mixtures are incubated in the usual way (37°C, 5% CO_2) for four days if using soluble immunogen, three days if using cell immunogens, before fusion (see Chapter 5).

Table 3.8

Immunogen	Dose/pretreatment
Soluble immunogen	1 μg or less (sterile solution)
Intact cells (able to survive in lymphocyte medium)	Mix two cultures
Intact cells (unable to survive in lymphocyte medium)	Fix cells first*
Intact cells (tumorigenic)	First suppress growth by gamma-irradiation (^{60}Co, 3300 rad)

*Fix adherent cells by incubating with sterile 0.25% glutaraldehyde for 20 min and wash repeatedly in sterile PBS before adding lymphocyte suspension.

If cells are not normally adherent try pretreating flasks with 1 mg poly-L-lysine/ml for one hour.

Chapter 4

Screening test design

4.1 INTRODUCTION

The notion that the selection of useful hybrid cells must be based upon a specific antibody-screening test has already been introduced. In this chapter, the requirements for such tests and their design in terms of hybridoma growth will be considered in some detail. The topic is dealt with before the chapter on fusion since it is important to establish the screening test before any hybridomas are generated. A selection of different assay formats in common usage is given in order to illustrate the important criteria.

The basic guidelines for any screening test which is to be used in hybridoma selection are that it must be specific for the antibody of choice, sufficiently sensitive and easily reproducible and capable of processing several hundred samples at a time. Time and effort spent at this stage could shorten the time taken to produce useful monoclonal antibodies, the final product being a direct result of the quality of the antibody-screening test.

In designing the test, it may be useful to adapt existing methods employing polyclonal antibodies. However, such methods are not necessarily directly transferable, in that the performance characteristics of the two systems may not be the same. Thus, although existing methods may act as a starting point for the design of a screening test, certain characteristics of hybridoma technology itself have to be considered and the needs that arise satisfied by the design of the final test format.

4.2 WHEN TO SCREEN

Firstly, the context in which the screening test is to be used should be considered, i.e. when and how often will the test need to be performed. Fig. 4.2 illustrates an idealised time course of hybridoma selection based on the expected products of one fusion handled by a single worker. Obviously, such a protocol will vary depending not only on actual results but also on available personnel and workload from other simultaneous fusions. However, in this typical protocol, the first antibody screen after fusion will be required when, by observation, the density of viable hybrid cells reaches approximately one third of the confluent level. This will usually occur between 10 and 14 days post-fusion, although longer periods may be required if the initial seeding density was low.

In the first screen there are likely to be 180 wells, each containing 150–200 μl of culture supernatant, up to 100 μl of which is available for use in the test. The percentage of cells that may be secreting specific antibody is highly variable, depending upon a number of factors which include the immunogenicity of the immunogen and the specificity of the screening test. On the basis of these first test results a number of wells are selected and the cells they contain transferred to larger (1.5-ml capacity) culture wells. The chosen cells are allowed to expand in number until they approach one third confluency (0.5×10^6 cells/ml). This expansion phase should take no longer than seven days but may be completed in a much shorter time if the cells show particularly vigorous growth. The cell population will then be ready for rescreening. It is advisable to rescreen at this stage because in this relatively early stage of selection the hybrid cells are particularly susceptible to chromosomal loss of antibody production or to the overgrowth by some vigorous non-secreting cells.

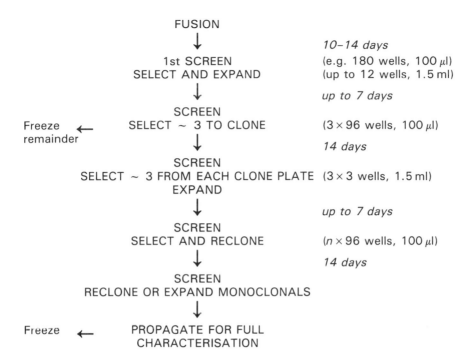

FUSION
↓   10–14 days
1st SCREEN   (e.g. 180 wells, 100 μl)
SELECT AND EXPAND   (up to 12 wells, 1.5 ml)
↓   up to 7 days
SCREEN
Freeze ←   SELECT ~ 3 TO CLONE   (3 × 96 wells, 100 μl)
remainder
↓   14 days
SCREEN
SELECT ~ 3 FROM EACH CLONE PLATE (3 × 3 wells, 1.5 ml)
EXPAND
↓   up to 7 days
SCREEN
SELECT AND RECLONE   (n × 96 wells, 100 μl)
↓   14 days
SCREEN
RECLONE OR EXPAND MONOCLONALS
↓
Freeze ←   PROPAGATE FOR FULL
CHARACTERISATION

Fig. 4.2 Timing of screening tests

Of the wells still shown to contain secreted antibody, a manageable number are selected for cloning, whilst the remainder are frozen for later retrieval if necessary. The most usual cloning method, that of limiting dilution (see Chapter 6), generates 96 new culture wells from each of the previous wells. Each clone plate will, therefore, yield $96 \times 100\,\mu l$ of culture supernatant for assay, within approximately two weeks. Statistically, only approximately 33% of the 96 wells will contain living cells, but it is rarely practical to select only these for testing. The screening test will need to accommodate all the 96 wells and also possibly several other clone plates at the same time. Of the wells positive for growth, an even smaller proportion, if any, will be secreting specific antibody. However, only those most strongly positive for antibody secretion, also displaying healthy growth, and preferably originating from one clone, should be selected for further development. It is possible, though unlikely, that a monoclonal cell line secreting specific antibody will be found at this stage, in which case only expansion and propagation of the cell line is required. Normally, several cycles of cloning and screening will be necessary before monoclonality is assured. A good indication of monoclonality is that in the final clone plate all wells displaying cell growth will also show specific antibody secretion.

4.3 TEST DESIGN

The purpose of the screening test is to select a group of cells secreting the desired antibodies out of the millions of irrelevant cells generated by the fusion process. The easiest way of locating these cells is to measure levels of the desired antibody that have been secreted into the culture medium. To facilitate this process, the cell mixture is distributed post-fusion into hundreds of individual culture wells. Alternatively, a less usual method is to grow cells in a semi-solid agar support so that particular cell colonies and their secreted products can be located.

In designing the screening test, it is essential to consider these and other growth characteristics of the living cultures. Established assay formats, that are already employing polyclonal antibodies, may be useful as a starting point, especially if existing equipment and reagents can be made use of, but the hybridoma growth characteristics cannot be ignored. The most important factors to consider can be categorised broadly under the headings of sensitivity, specificity, speed and scale.

4.3.1 Sensitivity

The antibodies to be measured are secreted into the media feeding living cell cultures. The cells will be growing rapidly (doubling times 12–24 hours) so their nutrient needs have to be met by replenishment of the medium. This requirement is in direct conflict with the desire to accumulate as much secreted antibody as possible. It would be sensible policy, therefore, to schedule the screening test for several days after a medium change, so that antibody levels are maximal, without compromising the nutritional requirements of the cells. The need to replenish the culture medium can be judged by colour change from orange to yellow as the pH becomes more acid. Obviously, the more sensitive the test, the less important is the need to allow time to build up sufficient antibody levels in the medium.

The concentration of specific antibody detected in a screening test needs to be compared and correlated with the degree of cell growth. This will include consideration of the total number of cells present, their growth rate, and the percentage viability. For instance, one particular well may contain a relatively small number of cells and be only moderately positive for antibody secretion. However, this well may eventually be preferable to one containing a larger number of cells giving a higher positive response in the screening test, since at this stage one will have no means of knowing what proportion of cells in any well are secreting specific antibody.

One also needs to consider whether the cells that produced the specific antibody are still flourishing. It is possible, especially in the early stages of development, that a new hybridoma cell line may discard excess chromosomes. Not only those coding for specific antibody are at risk, but also those necessary for vital functions. The experienced eye can distinguish between the original cell debris resulting from HAT selection and debris resulting from the recent death of a new colony. New mouse clones tend to grow as discrete colonies amongst the general cell debris, especially if left undisturbed by vigorous media changes. If such a colony is dying, it may indicate the loss of a vital chromosome, and the loss of the specific antibody-secreting clone.

In practice, it is often the case that the test is too sensitive and the problem becomes one of choosing which of the many positive wells to pursue. This could, of course, be due to lack of specificity in the test but could also be due to the not insignificant contribution, in the first post-fusion screen, of the specific antibody products of unfused dying spleen cells. The latter factor can be reduced to some extent by changing the culture medium frequently in the early stages.

The degree of sensitivity required of the screening test will depend to some extent on the proposed final application of the antibody being sought. For example, if the antibody is to be used in immunoassays, where high affinity is a requirement, a relatively insensitive screening test will enable one to concentrate on only those likely to be of higher affinity. Alternatively, if the required antibodies are to be used for immunopurification using affinity chromatography, lower affinity antibodies will be preferable which are more likely to be detected by a higher sensitivity screening test.

In the absence of any culture supernatant containing specific antibody, the initial screening test will be developed using a polyclonal antiserum derived from test bleeds of potential spleen donors (or preferably a larger pool from animals devoted to antiserum production). The titre of this antiserum is obviously highly variable but one would expect to detect antibody in undiluted culture supernatant at levels comparable to between 1 in 1000 to 1 in 10 000 of the antiserum.

4.3.2 Specificity

The need for specificity in the test is paramount, particularly when a relatively crude immunogen has been used. The non-specific reactivity detected by the test must also be low since the lower end of the detection range is more important in order to detect the low levels of antibody in culture media.

A test to detect non-specific immunoglobulin in the culture media is not usually useful for two reasons. Firstly, a good fusion will yield virtually 100% of wells containing growing cultures, a great many of which will be producing antibody. Since the aim of the screening test is to eliminate irrelevant cultures, this test is clearly of no use. Secondly, even if there were manageable numbers of cultures remaining after such a test, the results would be misleading. The highest immunoglobulin secretors may not necessarily be the highest specific immunoglobulin secretors.

In practice, even a specific antibody-screening test may reveal many cell culture wells with detectable antibody present. The number of antibody-secreting cultures may be such as to make it impossible for all the cultures to be taken to the next stage. In such circumstances, a selection has to be made, and only those wells offering the most promise should be pursued. It is often tempting, when starting hybridoma work, to culture too many different lines, with the result that time is wasted on poorer lines at the expense of the more valuable lines. Any cultures not selected in the first round need not be discarded, however, but may be stored frozen (see Chapter 5) and studied at a later, more convenient stage.

In some situations, it is difficult to attain a high level of specificity in the screening test with the materials that are available. For instance, it may be impossible to purify the antigen in sufficient quantities or it may even be the case that the precise nature of the antigen is unknown at this stage. In such circumstances, it is worth considering running a parallel screening test which is specific for one or more of the suspected major contaminants in the antigen preparation.

4.3.3 Speed

The time involved in performing the test is another important consideration. The time between sampling of culture supernatant and analysis of the result should not be more than a working day, or at least no more than 24 hours. The reason for this condition is based on the characteristics of hybridoma cell growth. As with myeloma cells, growth is very rapid, the doubling time usually being between 12 and 24 hours. Thus, as the cultures grow to fill their containers, any delay in expanding them will result in overcrowding, increasing the risk of the specific cell population dying out. Also, several days' delay in cloning a particular cell population will reduce the percentage of specific antibody secretors in that population, thus reducing the success of specific cloning.

In addition, the time involved in preparing the test should be minimal. For instance, the availability of reagents such as purified antigen, labelled antibody or antigen preparations must be assured by maintaining adequate and well-characterised stocks. The time required to prepare these stocks should not encroach on the time necessary to maintain the hybridoma cultures. Screening tests are often needed at short notice and are dependent on the vagaries in growth of the hybridoma cultures. Clearly, planning is extremely important, particularly if the screening test is based on antigen present on cultured cells, requiring co-ordination of the two cultures.

4.3.4 Scale

An important factor in developing a rapid screening test is the degree of automation that can be incorporated into it, such that the scale of operations does not become an intolerable burden. Many test formats can be relatively rapid when dealing with a few samples but in this situation there may be hundreds of samples to test at any time. For example, in the protocol suggested in the next chapter, the first screening test may include at least 180 samples and, subsequently, several clone plates of 96 wells each may need testing simultaneously. It is not usually practical to select wells out of a clone plate, to test only those containing cells (theoretically 33%). The process can be very time-consuming and liable to error.

Since cultures are often grown in 96-well plates, test design in a 96-well format would be useful to aid automation. There are many devices on the market for this purpose, from simple multipipetters to automatic reagent dispensers and even fully automated robots. Clearly, any screening test protocol that includes a labour-intensive or time-consuming step such as centrifugation, measurement of biological activity or the examination of tissue sections under a microscope will not be ideally suited to the current purpose.

If there is no alternative, it may be worth considering the possibility of distributing the fusion products between a smaller number of larger capacity wells. Thus, instead of the fusion products yielding 3×96 wells of 100 µl supernatant volume, 1 or 2×24 wells of 2 ml capacity may be set up. However, under these conditions, the sensitivity of the screening test may be more critical in that the antibody will have to be detected in larger volumes, and each well would contain a larger number of unwanted cells which would tend to reduce the efficiency of cloning.

4.4 CHOICE OF ASSAY

There are many different immunoassay formats in use to measure antibody and antigen but their adaptability to all the screening test requirements considered above is not always possible. The type of test chosen will, at least in part, depend on the characteristics of the antigen in question. It would be an advantage to design a screening test which detects antibody in a format compatible with the antibody's proposed final use. For instance, if the monoclonal antibody is required to detect the biological activity of the antigen, the use of inactivated or denatured antigen in the screening test may not reveal and/or distinguish the relevant antibody. Similarly, a monoclonal antibody required for the recognition of live cells may not be picked out by the use of fixed cells in the screening test. A monoclonal antibody required to identify dissociated antigen subunits in polyacrylamide gels containing sodium dodecyl sulphate may be missed if whole undissociated antigen is used in the screening test. Any antibody required for therapeutic applications in vivo may be required to fix complement. Consequently, a cytotoxic screen in which complement is fixed might be appropriate. Similarly, if the antibody is to be used in affinity chromatography using protein A, the use of protein A binding in the screening test might be useful.

It should always be borne in mind that, unlike multivalent polyclonal antibody preparations, an individual monoclonal antibody detected by one type of screening test may not necessarily function when applied in a different type of assay. For instance, one fairly common drawback has been the poor response of monoclonal antibodies applied as probes in immunoelectrophoresis, whereas they may be ideal in immunoassays of IRMA or ELISA formats. If necessary, the search for an antibody reagent of universal application can be aided by introducing more than one type of screening test in the early stages.

The following sections will consider in turn different types of immunoassay that have been used as hybridoma-screening tests, with practical examples that may be adapted to suit the particular needs of a given antigen. Most of these assays can be categorised on the basis of how the antigen–antibody complex is separated from non-reactants, i.e. by being bound to a solid phase or removed from a liquid phase. Various labels can then be applied to detect the presence of specific antibody. The relative merits of these assays will be considered in terms of the ideal criteria discussed above.

4.4.1 Solid-phase assays

4.4.1.1 ENZYME-LINKED IMMUNOSORBENT ASSAY (ELISA)

Probably the most commonly used assay adapted for hybridoma screening is the ELISA. It can satisfy all the criteria demanded by hybridoma growth and in terms of ease of use and amenity to automation is far superior to most other systems. These factors can be best explained by referring to the specific examples given below. Most readers will be familiar with the basic principles of ELISAs, but detailed information on all aspects of enzyme assays can be found in an excellent book by Tijssen (1985). In this context, the presence of an antibody in culture supernatant may be measured by its binding to specific antigen attached either directly or indirectly to the wells of a 96-well plate. The antibody binding is quantified by adding the relevant anti-species immunoglobulin to which an enzyme is bound, followed by a chromogenic substrate to that enzyme. Two illustrative general ELISA procedures are described in Protocols 4.4.1.1a and 4.4.1.1b. By including appropriate standard reagents, the amount of antibody present can be quantified either visually or more accurately in spectrophotometers designed to read 96-well plates, some of which include computer analysis.

**PROTOCOL 4.4.1.1a: ELISA SCREENING TEST FOR ANTIBODY
IN CULTURE SUPERNATANTS (direct soluble antigen
adsorption to assay plate)**

For more detailed discussion of the practical aspects of each step in ELISA, see
section 4.4.1.1.1.

Materials:

    Microtitre plates for ELISA
    Coating buffer
    Wash buffer
    Anti-mouse or rat Ig–enzyme conjugate
    Enzyme substrate
    Substrate stopping solution
    Blocking solution
    Test samples (culture supernatant)
    Positive control sample (e.g. several dilutions of polyclonal antiserum)
    Negative control samples (e.g. culture medium, non-immune serum, irrelevant
      antibody)
    Multichannel pipette
    Plate washer or wash bottle

Method:

1. Adsorb predetermined maximal concentration of antigen to the microtitre plate.
   Dilute in coating buffer and incubate $100\,\mu l$ per well overnight at 4°C. It is
   advisable to include negative controls here, particularly in the developmental
   stages, such as wells coated with irrelevant protein (i.e. 1% BSA in coating
   buffer).
2. Aspirate antigen solution and block unbound sites on the plastic by incubating
   $100\,\mu l$ of blocking solution per well for 15 min.
3. Wash off unbound protein three times with wash buffer.
4. Add test samples of culture supernatant. Incubate $100\,\mu l$ per well for one hour
   at ambient temperature. Include some wells with known positive antibody and
   known negative antibody.
5. Wash off unbound antibody three times with wash buffer.
6. Add enzyme-labelled anti-mouse Ig (or rat or human as appropriate) at the
   recommended or predetermined dilution. Incubate $100\,\mu l$ per well for one hour
   at ambient temperature.
7. Wash off unbound label three times with wash buffer.
8. Add enzyme substrate. Incubate $100\,\mu l$ per well at ambient temperature until
   colour develops sufficiently (ideally approx. 30 min).
9. Do not wash off. Add $50\,\mu l$ of the appropriate stopping solution and assess either
   visually or quantitatively in a spectrophotometer designed to read microtitre
   plates.

continued on next page

continued

Suitability as hybridoma screening test:

Sensitivity: Potentially good.

Specificity: Dependent on purity of coating antigen. If complete purification is possible, the relevant epitope may be destroyed in the process or in binding to the plate.

Speed: More than adequate. Maximum time between addition of culture supernatant and result, 2.5 hours, and could be reduced even further.

Scale: 96-well format allows easy handling of hundreds of samples simultaneously. Various degrees of automation are possible.

Reagent availability: Purifying antigen may be the biggest problem. Also as much as 100 μg of pure antigen may be needed per plate. Labels are stable, relatively cheap and easily available commercially.

Hardware: Quantifying the result can take seconds per plate, with spectrophotometers designed to read 96-well plates, but since the results are visual a qualitative assessment can be made by eye.

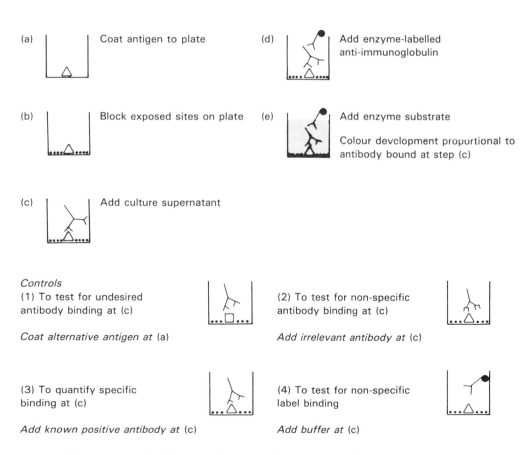

(a) Coat antigen to plate

(b) Block exposed sites on plate

(c) Add culture supernatant

(d) Add enzyme-labelled anti-immunoglobulin

(e) Add enzyme substrate
Colour development proportional to antibody bound at step (c)

Controls
(1) To test for undesired antibody binding at (c)

Coat alternative antigen at (a)

(2) To test for non-specific antibody binding at (c)

Add irrelevant antibody at (c)

(3) To quantify specific binding at (c)

Add known positive antibody at (c)

(4) To test for non-specific label binding

Add buffer at (c)

Fig. 4.4.1.1a ELISA screening test: direct antigen adsorption to assay plate

PROTOCOL 4.4.1.1b: ELISA SCREENING TEST FOR ANTIBODY IN CULTURE SUPERNATANTS (indirect soluble antigen adsorption to assay plate)

See also section 4.4.1.1.1.

Materials:

  As for Protocol 4.4.1.1a
  Antibody for coating (Ig fraction, preferably affinity purified)
  Antigen solution for step 4 (purification not necessary)

Method:

1. Adsorb predetermined maximal concentration of antibody to the microtitre plate. Dilute in coating buffer and incubate 100 μl per well overnight at 4°C. Include negative controls of irrelevant antibody diluted in coating buffer.
2. Aspirate antibody solution and block unbound sites on the plastic by incubating 100 μl of blocking solution per well for 15 min.
3. Wash off unbound protein three times with wash buffer.
4. Add excess antigen diluted in wash buffer. Incubate 100 μl per well for one hour at ambient temperature. Include a negative control of a few wells of irrelevant protein.
5. Continue as in Protocol 4.4.1.1a steps 3–9 with serial incubations of test culture supernatant, second antibody–enzyme conjugate, enzyme substrate and stopping solution.

Comments:

Advantages over previous format as hybridoma screening test:
  The specificity is improved and is now dependent on the purity of the coating antibody. An Ig preparation rather than an antiserum will bind to the plate more effectively. The antigen need not be pure; for example, a serum protein could be added as a crude serum.

Disadvantages:

  Since the antigen is anchored by an antibody, many epitopes may already be occupied, so that the test 'monoclonal' antibody has no site to bind to. This is more of a problem with small antigen molecules but not with large multimeric antigens. However, it could be turned to advantage, in seeking different monoclonal antibodies. If an already established monoclonal antibody is used to anchor antigen and therefore block one epitope, any test antibody is likely to be directed at a different epitope.

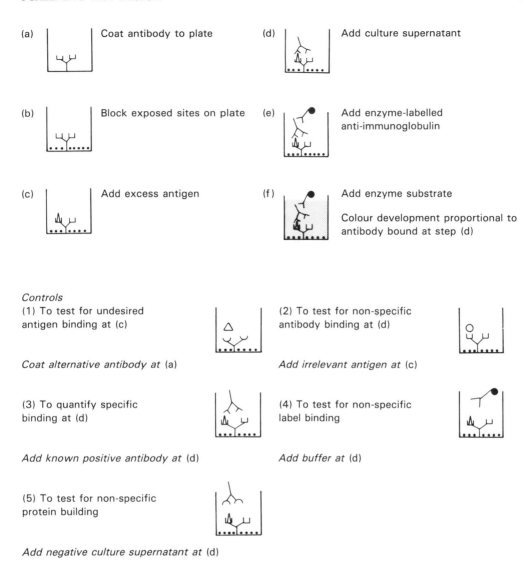

(a) Coat antibody to plate

(b) Block exposed sites on plate

(c) Add excess antigen

(d) Add culture supernatant

(e) Add enzyme-labelled anti-immunoglobulin

(f) Add enzyme substrate

Colour development proportional to antibody bound at step (d)

Controls
(1) To test for undesired antigen binding at (c)

Coat alternative antibody at (a)

(2) To test for non-specific antibody binding at (d)

Add irrelevant antigen at (c)

(3) To quantify specific binding at (d)

Add known positive antibody at (d)

(4) To test for non-specific label binding

Add buffer at (d)

(5) To test for non-specific protein building

Add negative culture supernatant at (d)

Fig. 4.4.1.1b ELISA screening test: indirect antigen adsorption to assay plate

4.4.1.1.1 Practical considerations of ELISA screening test design

For clarification the following considerations refer specifically to ELISA, but many of the points can be applied to other solid-phase assay systems. Each stage of the assay is considered in turn, i.e. the solid phase, antigen binding, test antibody binding, controls, enzyme conjugates and their substrates.

Solid phase

The solid phase used in ELISAs can vary in form and composition but the most convenient for the present purpose are 96-well flat-bottomed microtitre plates made of polystyrene or polyvinylchloride, treated commercially such that their binding characteristics are suitable for ELISA. Although the quality of binding to these plates (high specific, low non-specific, uniformity between wells) has improved greatly over the years, there can still be considerable variation between manufacturers and even between batches from the same manufacturer. So, each test should be developed on the most suitable batch of plates, which should not be changed without also re-establishing the ideal conditions. It cannot be assumed that the assay performance will be the same on two different batches of plates.

Antigen binding

(a) Soluble protein   Concentrations of between 1 and 10 μg/ml are usually sufficient to give maximal binding to the plate, but the precise concentration will have to be established for each protein by serial dilution. Too little will reduce the sensitivity of the test and too much, besides being wasteful, results in the formation of multiple layers so that binding is unstable during washing procedures. For best results the protein should be as pure as possible, not only to improve the specificity of the test, but also because protein mixtures have complex binding characteristics.

The most widely used coating buffer is 50 mM carbonate, pH 9.6 (see Appendix A), but 10 mM Tris-HCl, pH 8.5, containing 100 mM NaCl, or 10 mM sodium phosphate buffer, pH 7.2, containing 100 mM NaCl, are also used. In the initial development of the test, several different buffers should be tried. Tween 20, a detergent included in the wash buffer, should never be included in the coating buffer since it will interfere with protein binding. Normal incubation conditions are overnight at 4°C, but this time can be shortened if the temperature is increased to 37°C or the protein concentration is increased. Plates should be covered and sealed to prevent evaporation. It is possible, in some circumstances, to keep coated plates at 4°C for several weeks before use, although the exact conditions should be well established before use.

(b) Cells   If the antigen of interest is a cell membrane component, the cells themselves can be attached to the plastic surface via a bridging molecule, which has a high affinity for the plastic. Suitable bridges include glutaraldehyde (0.25%), poly-L-lysine (10 μg/ml), or phytohaemagglutinin-P (PHA, 20 μg/ml), and also antibodies but these can present special problems in a hybridoma-screening test as explained above. Assay plates are pretreated with the bridging molecules at these concentrations in PBS for one hour at 37°C, before incubation with $1–5 \times 10^6$ well-washed cells/ml, with frequent shaking to distribute the cell layer. The plates are centrifuged at 100 g for 3 min. If the plates are to be stored they should be fixed by immersion in a one-litre beaker of 0.25% glutaraldehyde in PBS at 4°C for 5 min (without air trapping). They should then be washed gently in PBS before adding a blocking agent to saturate unreacted sites, i.e. 1% BSA in PBS for one hour or an egg white solution (stir chicken egg white in PBS for one hour, centrifuge at 10 000 g for 15 min) for PHA-coated plates.

Similar bridging molecules can also be used to bind soluble antigens that are poorly adsorbed to plastic, such as small oligopeptides and haptens. Details of individual methods and references can be found in Tijssen (1985).

(c) Non-specific binding   In developing any assay or screening test it should be firmly established that none of the subsequent layers of the sandwich bind directly to the plate in the absence of antigen. Non-specific binding of this kind is not always a problem, but is more likely if cells or complex proteins are used as coating antigen. The best way of reducing this effect is to block the remaining binding sites on the plastic, by incubating the coated plate for 15 min with 1% BSA, gelatin, ovalbumin, casein or milk powder. The addition of the detergent Tween 20 (0.05%) in the wash buffer also helps to remove unbound or weakly adsorbed molecules without interfering with antigen–antibody reactions. If test samples are culture supernatants accumulated from cultures of dying cells and general cellular waste, i.e. if the medium has not been changed between fusion and first screen, it is possible that the debris will contribute to non-specific binding. This effect can be minimised by changing the culture medium a few days before screening.

If the antigen is bound indirectly to the plate by an antibody as shown in Protocol 4.4.1.1b, ensure that there is no cross-reactivity between the anti-species immunoglobulin–enzyme conjugate and the coating antibody, even if the antibodies are from different species. If alternative antibodies are not available, try coating with Fab fragments of antibody. If negative controls are included at each stage of the test, any non-specific binding can be monitored and corrected before time is wasted following false positives.

Inefficient washing of the wells between incubation steps is a common cause of non-specific variability. Wash by hand, by filling and aspirating the wells at least three times with PBS-Tween 20 (use a wash bottle to fill and invert and shake the plates to empty, avoiding air traps), or use one of the automatic plate washers available commercially.

Test antibody binding and controls

When culture supernatants are incubated with the plate, at least one well per plate should contain fresh culture medium as negative control or ideally culture supernatant from cells secreting antibody totally unrelated to the specificity being sought. As positive control, serial dilutions of polyclonal antiserum from the same species or established monoclonal antibody should also be included together with similar dilutions of non-immune serum. Then, not only can results be compared and quantified between assays but also assay variability can be monitored. In the event of no specific antibody-secreting cells being detected, a positive control will enable one to assess whether it is a true result or due to assay failure.

Antibody levels in culture supernatant rarely exceed $10\,\mu g/ml$ so they are normally tested without dilution. Most of the wells screened will be negative for specific antibody so the mean value of these can represent background levels. Specific antibody levels can be expected to be at least double or even treble the background. Background levels are likely to increase as the incubation period increases. Most specific binding will occur within one hour of incubation at room temperature. Shorter incubation times can be tolerated at higher temperatures but greater sample variability is likely.

Enzyme conjugates and their substrates

The binding of test and standard antibody is quantified by adding 'second antibody' or species-specific anti-immunoglobulin antibody directed against the Fc region of the test antibody. The second antibody is pre-conjugated to an appropriate enzyme, instructions for which are given in Chapter 7, but it is more usual to obtain the conjugate commercially, given the reasonable cost and convenience.

Second antibodies can be selected for their specificity against immunoglobulin in general, against the separate classes, IgG or IgM or even against different subclasses. The anti-subclass reagents are more expensive, but if only a particular antibody subclass is required, the use of these reagents in the early stages will save time.

With few exceptions, the preferred enzyme used in ELISA is horseradish peroxidase (HRP), which with its substrate *o*-phenylenediamine (OPD) can produce the highest signal-to-noise ratio. An orange colour is produced which, in low concentrations, can be measured at 492 nm. One drawback is that OPD is potentially mutagenic but there are alternative substrates that are less hazardous. Beware of HRP inactivation if incubation temperatures above 15°C are used, although addition of Tween 20 will delay inactivation. HRP conjugates are not advisable when using certain cellular material or plant tissue as the antigen, since endogenous peroxidases will interfere with the reaction. In these situations, use alternative enzymes such as alkaline phosphatase (AP) with its substrate, *p*-nitrophenyl phosphate (p-NPP). AP, however, is particularly abundant in animal tissues involved in nutrient transport, developing tissues and secretory organs. If AP is the chosen conjugate, the wash buffer should be changed to 3 M Tris-HCl, pH 8.0, since PBS contains sufficient inorganic phosphate, even after washing, to inhibit AP.

Manufacturers usually recommend an enzyme conjugate dilution of 1 : 1000 but often this can be diluted further. Ideally, colour development up to two units of absorbance should take 30 min. A faster time than this leads to unacceptable variation between samples and indicates the need to dilute the conjugate. All colour development is a continuous process, so the reaction needs to be stopped before the colour becomes too intense to be quantified. Also OPD is light sensitive so longer colour development times will result in higher backgrounds.

4.4.1.2 IMMUNODOTS

If obtaining enough antigenic material to bind to a solid phase presents a problem, it might be worth considering immunodots. Very small amounts of antigen (1 μl containing 100 pg or less in a diameter of less than 1 mm) and as few as 3000 cells per sample can be absorbed onto nitrocellulose membranes (in a 96-well format if desired). Proteins, nucleic acids, cell membranes, subcellular organelles, fungi, protozoa, bacteria and viruses have all been assayed in this way (Hawkes et al, 1982). The method also has the advantage of enabling detergent-solubilised membrane antigens to be bound to the nitrocellulose. If the concentration of non-ionic detergents (Triton X-100, Tween 80) is greater than 0.01%, a fixation step will also be required after dotting, involving immersion for 15 min in a solution of 10% acetic acid and 25% isopropanol followed by rinsing in water. If the sample contains SDS, a seven-fold excess of Triton is beneficial. Non-specific binding sites on the membrane can be blocked with irrelevant protein or simply by including Tween 20 at 0.05% in reagent diluents. Subsequent detection of test antibody binding can be with anti-immunoglobulin labelled with enzyme or radiolabel. Unlike ELISA, the enzyme substrate should yield an insoluble coloured product, such as would be produced with HRP and 4-chloro-1-naphthol or AP and 5-bromo-4-chloro-3-indolyl phosphate (BCIP) and nitroblue tetrazolium (NBT) (see Appendix A).

Immunodot assays can also be performed on diazobenzyloxymethyl (DBM) paper, which is cheaper and easier to handle than nitrocellulose (Motta & Locker, 1986).

4.4.1.3 IMMUNORADIOMETRIC ASSAY (IRMA)

The two ELISA formats described in Protocols 4.4.1.1a and 4.4.1.1b can equally well be adapted for use with radiolabels such as ^{125}I, with the advantage that sensitivity can be increased at least 10-fold. This extra sensitivity, however, may not always be necessary in practice. Fig. 4.4.1.3 illustrates the results of two screening tests performed on the same samples of culture supernatants, 14 days after fusion. The format of the assay was the same in each test, i.e. coating antibody–antigen–test supernatant–labelled antibody. The difference between the two tests was that the final detection label was a radiolabel in one and an enzyme label in the other. Despite the increased sensitivity of the IRMA over the ELISA, the end result was that only two extra samples were selected for further development, ten from IRMA and eight from ELISA.

Polystyrene tubes are the usual solid phase in IRMAs but automation can be improved for hybridoma screening by replacing these with 96-well plates in one of two ways. The problem lies in the need to count wells individually in a gamma-counter at the end of the test. Individual wells can be bought set in a 96-well tray so that at the end of the assay they can be removed and placed in tubes ready for counting. Alternatively, the solid phase could be a flexible PVC 96-well plate which can be cut up into individual wells with scissors at the end of the assay and counted separately.

The time taken to perform the test would be the same as for ELISA, but counting the samples would take longer, even allowing for 30 seconds per sample. Nevertheless, the overall time falls within the upper limit of 24 hours set by the hybridoma growth characteristics.

The biggest disadvantage in using IRMAs as screening tests lies with the label. It involves the use of radioactivity, with all its risks to health and safety, but radioactive labels are unstable, requiring replacement approximately every three weeks. They are expensive to buy, but time-consuming to make, requiring lengthy purification of antiserum and evaluation of each new product. Details of antibody purification and radiolabelling by various methods are given in Chapter 7.

4.4.1.4 IMMUNOHISTOCHEMISTRY

It may be necessary to select monoclonal antibodies that recognise cell types or subcellular components that cannot be isolated easily for use in the assay formats described above. In this case, it could be extremely useful to observe the binding of antibody to the cells or tissue in situ, i.e. by adapting an immunohistochemical assay.

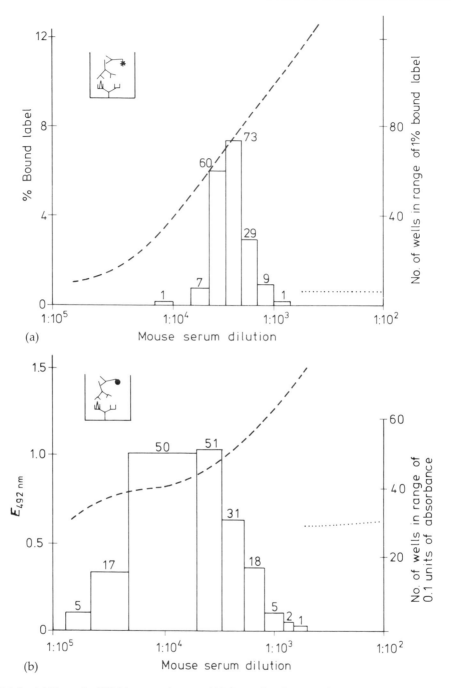

Fig. 4.4.1.3 (a) Two-site IRMA screening test (14 days after fusion). (b) Two-site ELISA screening test (14 days after fusion). Standard curves of % bound label absorbance obtained from polyclonal mouse antiserum. Histogram superimposed of numbers of culture wells resulting in a given % bound label absorbance. Key: ----, polyclonal mouse antiserum; ·····, non-immune mouse serum. Results: (a) Ten good positive wells selected for further culture and cloning + 29 weaker positives. These samples were the same as those tested in the two-site ELISA; however, despite the greater sensitivity of the IRMA, only two extra positive wells are revealed. (b) Eight good positive wells selected for further culture and cloning + 18 weaker positives. N.B. The background levels were unusually high in this test but inclusion of appropriate controls enables clear positives to be identified

The use of tissue sections can be limiting because of the need to observe possibly hundreds of individual slides, the result of which is bound to be subjective. However, this can be a useful secondary test, when a more practical primary test has selected a manageable number of samples.

The problem of processing so many samples can be overcome with primary cell cultures or cell lines, by changing their normal growth containers. For instance, they can be grown on sterile glass plates which have been marked on the underside with a 96-well grid or on the plastic lids of 96-well tissue culture dishes (those with condensation rings provide 96 defined culture areas). Cell attachment to plastic lids can be promoted by sulphonation (pretreat lids with concentrated H_2SO_4 for 10 min, and wash thoroughly with sterile water until pH is neutral). The plate is then immersed in a large covered Petri dish with medium for the growth phase and, at the appropriate stage, washed and fixed if necessary for the immunohistochemical phase.

The possibility of storing the test cultures after fixation should be established. Given the rapid growth of hybridomas, it would be virtually impossible to co-ordinate two different live cultures so that the test cells are ready for assay when the hybridoma cells are at the right stage. Non-adherent cells can be fixed by immersion of the plate in 4% formaldehyde in PBS for 10 min. In all assays using live cells it is advisable to keep the cells cool throughout in order to avoid antibody-induced capping, endocytosis or shedding of antigen–antibody complexes during the procedures. Only 10 μl of culture supernatant and controls need be added per cell sample. The second antibody label could be enzymatic or fluorescent. Although generation of the finished, labelled samples can be fairly rapid and well automated, the need to observe each sample microscopically remains the limiting step.

4.4.2 Liquid-phase assays

4.4.2.1 RADIOIMMUNOASSAY (RIA)

The simplest form of radioimmunoassay involves the incubation of radiolabelled antigen with test antibody and the subsequent separation of the insoluble immune complex from the soluble mixture of unreacted molecules. For several reasons, this format would be unsuitable for the detection of monoclonal antibodies. Firstly, the separation of immune complex usually requires centrifugation or filtration which necessarily involve individual sample handling. Secondly, the small immune complex formed between monoclonal antibody and antigen is highly unlikely to precipitate spontaneously. Thirdly, labelling of the antigen is not always possible, particularly if it is a complex molecule or bound to cell membrane and difficult to purify. Fourthly, one must consider the disadvantages in using radioactivity, as described earlier, namely the safety hazards and instability of the label.

Obviously, the format can be altered to accommodate some of these objections. For instance, an excess of unlabelled antigen can be allowed to react with the putative monoclonal antibody in solution. The complex can then be precipitated by the addition of radiolabelled anti-mouse (or rat or human) immunoglobulin (see Fig. 4.4.2.1). The labelled second antibody could even be coated to magnetic beads so that the centrifugation step could be substituted by magnetic separation (Kemshead & Ugelstad, 1985). The specificity of the test is dependent on the purity of antigen but, if this is

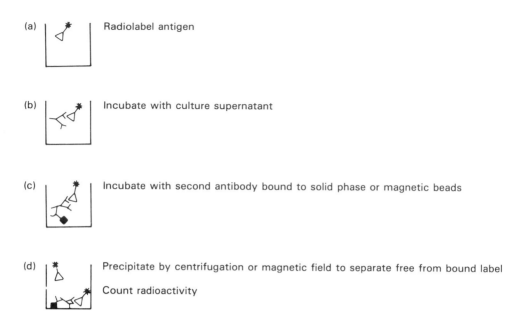

(a)   Radiolabel antigen

(b)   Incubate with culture supernatant

(c)   Incubate with second antibody bound to solid phase or magnetic beads

(d)   Precipitate by centrifugation or magnetic field to separate free from bound label

        Count radioactivity

Fig. 4.4.2.1 RIA screening test

not a problem, RIA may be able to detect antibodies that recognise antigen in its natural free state unaltered by binding to a solid phase. This is an important consideration if the final application of the antibody is to be in this type of assay.

4.4.3 Cytolysis and agglutination assays

A cytolytic screening test was originally used by Kohler and Milstein in their early monoclonal antibody work. It was a modified plaque assay, in which sheep red blood cells (SRBC) and a source of complement were overlaid on an agar plate containing the growing hybridoma clones. The secretion of anti-SRBC by specific clones was indicated by an area of SRBC lysis around those clones. The application of haemolytic assays can be widened by binding specific antigens to the red cell surface (see below) and haemolysis can be observed qualitatively in 96-well microtitre plates when culture supernatants containing the relevant antibody are added (see Fig. 4.4.3a). Sensitivity is not as good as with the quantitative assays described above, although lysis can be quantitated by measuring the red colour released spectrophotometrically at 412 nm or the release of ^{51}Cr previously incorporated into the cells. Other cell types can be used instead of SRBCs, where the antigen of interest is a component of the cell surface membrane, but, in the absence of the natural colour indicator, a viability test using nigrosin dye would be necessary. In the interests of sensitivity, such tests are not the first choice for hybridoma screening. In addition, antibodies that might recognise cell surface epitopes but were not reactive with complement would not be revealed.

Immunoassays based on haemagglutination have been used extensively in blood group analysis. Exogenous soluble antigens can be absorbed onto the cell surface by various

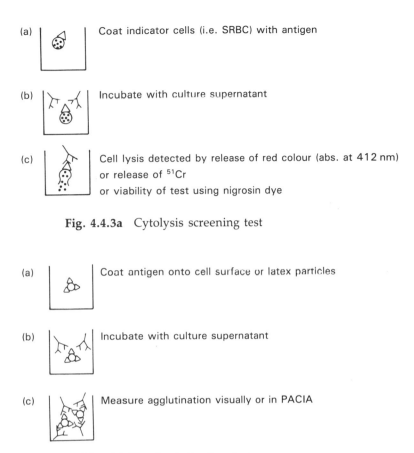

(a) Coat indicator cells (i.e. SRBC) with antigen

(b) Incubate with culture supernatant

(c) Cell lysis detected by release of red colour (abs. at 412 nm)
or release of ^{51}Cr
or viability of test using nigrosin dye

Fig. 4.4.3a Cytolysis screening test

(a) Coat antigen onto cell surface or latex particles

(b) Incubate with culture supernatant

(c) Measure agglutination visually or in PACIA

Fig. 4.4.3b Agglutination test

means (see Chapter 5). On mixing with hybridoma culture supernatant, the presence of specific antibody will cause the antigen-coated SRBCs to agglutinate, which is measured visually in round-bottomed microtitre plates (see Fig. 4.4.3b). The degree of cell agglutination is directly proportional to the amount of antibody present. Haemagglutination can be a very quick test, but sensitivity may not be adequate and the antigen-coated SRBCs have a shelf life of only two to three weeks. Latex particles of $0.8\,\mu m$ diameter, to which antigens can be coupled with carbodiimide, have been used as red cell substitutes in agglutination assays. They are more stable and also give fewer non-specific reactions. Agglutination can be quantified by sophisticated instruments such as PACIA (Technicon).

4.4.4 Biological assays

If the desired monoclonal antibody is to be used to study the biological activity of a particular antigen, the sooner the antibody is tested in such an assay, the better. Most biological assays will involve individual sample handling and may require sample volumes greater than $100\,\mu l$, and so are not ideal. Consider using such a test as a secondary evaluation after the first screen or reduce the initial numbers of cultures and increase the volumes and also recruit extra pairs of hands if necessary.

Functional assays should be interpreted with caution, since results may be affected by contaminants such as mycoplasma in the culture supernatant. Tests based on modulation of lymphocyte function, or the incorporation of [^{3}H]thymidine, are known to be affected by mycoplasma contamination.

4.5 CONCLUSION

The foregoing recommendations are not intended to be exhaustive but should highlight those factors that each experimenter must consider when applying his or her own knowledge of a particular antigen to the design of a useful screening test which most closely satisfies the criteria of sensitivity, specificity, speed and scale defined at the outset. The timing of selection procedures is crucial to the successful production of monoclonal antibodies with the minimum of unnecessary production work. Unproductive work, following cell lines that are ultimately unusable, can be minimised if it is always borne in mind that quality is more important than quantity.

Chapter 5

Myeloma cells, lymphocytes, fusion

5.1 MYELOMA CELLS

5.1.1 Origin

During the 1960s, methods for the production of large quantities of antibody for structural and genetic study were sought. The main source of antibody to emerge was the culture of multiple myeloma cells. Myeloma or plasmacytoma is a neoplasm of antibody-producing cells, each tumour representing the proliferation of a single clone of antibody-forming cells. Because of their ready availability a large number of murine myelomas have been generated and studied. In addition to providing information on the structure, biosynthesis and genetics of immunoglobulins, a number have been adapted to continuous survival in cell culture.

The myeloma system alone has been disappointing as a source of antibodies with known antigen binding specificity and it has not proved possible to generate antigen-specific myelomas. However, the fusion of myelomas with specific antibody-producing lymphocytes has proven extremely useful in the generation of antibodies with predetermined specificity. Of the cell lines adapted for continuous survival in culture the most popular ones for murine hybridoma production are those descended from the MOPC-21 myeloma.

5.1.2 Post-fusion selection criteria

Cell fusion for the immortalisation of spleen cells requires a means of selection for mixed hybrids (myeloma–spleen cell hybrids). The favoured selection procedure depends on the capacity of cells to use the 'salvage' pathway for guanosine production when the main biosynthetic pathway is blocked (usually by the antibiotics aminopterin or amethopterin). This salvage pathway relies on the presence of the enzyme hypoxanthine guanine phosphoribosyl transferase (HGPRT). Cells lacking the enzyme will die when grown in media containing hypoxanthine, aminopterin and thymidine because both pathways necessary for the formation of the purine precursors of DNA are blocked. However, a cell which is HGPRT$^-$ will grow in the selection medium if it is fused with a cell which is HGPRT$^+$. Myelomas which are HGPRT$^-$ are needed for successful selection of myeloma (HGPRT$^-$) and spleen cell (HGPRT$^+$) hybrids in the hypoxanthine–aminopterine–thymidine (HAT) medium.

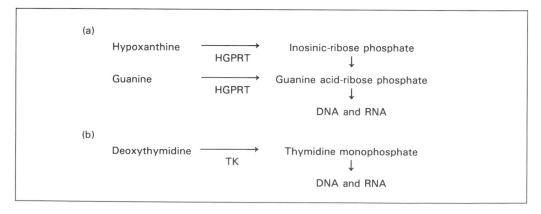

Fig. 5.1.2   (a) Purine biosynthesis salvage pathway. (b) Pyrimidine biosynthesis salvage pathway

Table 5.1.2 Post-fusion selection

Cell type	DNA synthesis		Survival in HAT medium
	Salvage pathway	De novo pathway	
Myeloma	HGPRT⁻	Aminopterin sensitive	Die (No DNA synthesis)
Spleen	HGPRT⁺	Aminopterin sensitive	Die (Finite survival in vitro)
Myeloma–spleen hybrid	HGPRT⁺	Aminopterin sensitive	Live
Myeloma–myeloma hybrid	HGPRT⁻	Aminopterin sensitive	Die (No DNA synthesis)
Spleen–spleen hybrid	HGPRT⁺	Aminopterin sensitive	Die (Finite survival in vitro)

The isolation of HGPRT⁻ myeloma cells is relatively easy because the enzyme is coded for on the single active X chromosome present in each cell. This means that only a single mutation is required to result in the loss of HGPRT. Selection of HGPRT⁻ cells is carried out in the presence of the toxic base analogues, 8-azaguanine (8-AG) or 6-thioguanine (6-TG), which are incorporated into DNA via HGPRT. It is not usually necessary to select for enzyme deficiency in myeloma cells since there is a range of useful HGPRT⁻ cell lines available commercially (see Table 5.1.3). In situations where it might be necessary, cells should be exposed to increasing concentrations of inhibitor (1–100 μg/ml) over six to eight weeks with several cloning steps, selecting for the most vigorous growth and least reversion to the normal state. Thioguanine is a more reliable selection agent than azaguanine (Evans & Vijayalaxmi, 1981).

Alternatively, myeloma cells can be selected to be deficient in the enzyme thymidine kinase (TK), by growth in bromodeoxyuridine. The substrate for TK is thymidine in the pyrimidine salvage pathway. Natural ouabain sensitivity is a property of use, particularly in the selection of heterohybridomas of human/mouse origin. Human cell lines normally die in the presence of ouabain at 10^{-7} M, whereas rodent lines are resistant up to 10^{-3} M. Unfused human cells can therefore be selected against.

Originally, the myeloma cell lines from which HGPRT⁻ variants were selected were also characterised by their capacity to secrete immunoglobulin molecules of their own (P3-X63-Ag8). Clearly, for monoclonal antibody production, this is undesirable since it reduces the proportion of hybrids which potentially will secrete wholly lymphocyte-derived antibodies. One of the most popular fusion partners derived from such early murine myeloma lines was P3/NS-1/1-Ag4-1. This myeloma produced only kappa light chains of immunoglobulin which were only secreted along with lymphocyte immunoglobulins following hybridisation.

A non-immunoglobulin-producing mouse myeloma cell line was isolated for the first time by Schulman et al (1978). The line, known as Sp2/0-Ag14, showed a variable efficiency of fusion although it was recognised as a potentially useful partner for generating hybridomas making truly monoclonal antibodies. Kearney et al (1979) identified another mouse myeloma cell line (X63-Ag8-653) that had lost the capacity for immunoglobulin expression but which still permitted the formation of antibody-secreting hybrid cell lines. Such 'non-producing' myeloma partners for hybrid production require rigorous selection in order to isolate them from the parent strains. Of the 'non-producer' mouse myeloma lines the latter two have become the most popular for the production of useful hybridomas although other lines have since been identified, i.e. NSO/1 (Galfre & Milstein, 1981). Most of the useful rodent-derived myeloma cell lines are available commercially and are listed below.

5.1.3   Rat myeloma cells

Similarly, a range of cell lines of rat origin is also available. Y3-Ag.1.2.3. (Y3) (Galfre et al, 1979) is an azaguanine-resistant line which secretes kappa chains whereas YB2/O (Galfre & Milstein, 1981) is totally incapable of immunoglobulin production and 1R983F (Bazin, 1982) has been proposed as a non-producer line suitable for hybridoma production. The maintenance of these cell lines is essentially similar to that of the mouse lines.

5.1.4   Human 'myeloma' cells

The search for suitable human myeloma cell lines has been extensive but hampered by other problems in producing human hybridomas, not least of which is the source of highly sensitised immune lymphocytes (see Chapter 8). Nevertheless, true human myelomas are notoriously difficult to establish in tissue culture. They are usually derived

Table 5.1.3 A selection of available rodent myeloma cell lines

Cell line	Abbreviation	Derivation	Ig secretion H	L	Reference
Mouse					
P3-X63-Ag8	P3	Balb/C mouse myeloma MOPC21	IgG1	$\varkappa$	Kohler & Milstein (1975)
P3/NS-1/1-Ag4-1	NS-1	P3	—	$\varkappa$	Kohler & Milstein (1976)
Sp2/0-Ag14	Sp2	P3/spleen cell hybrid	—	—	Shulman et al (1978)
P3-X63-Ag8.653	653	P3	—	—	Kearney et al (1979)
Rat					
210.RC43.Ag1		Lou/C rat myeloma	—	$\varkappa$	Cotton & Milstein (1973)
Y3-Ag.1.2.3.	Y3	Lou/C rat myeloma R210.RCY3	—	$\varkappa$	Galfre et al (1979)
YB2/3.0 Ag20	Y0	Y3/AO rat spleen hybrid	—	—	Galfre et al (1981) Kilmartin et al (1982)
IR983F	983	Lou/C rat myeloma	—	—	Bazin (1982)

Cell lines available in UK and Europe from ECACC, and in the USA from ATCC.

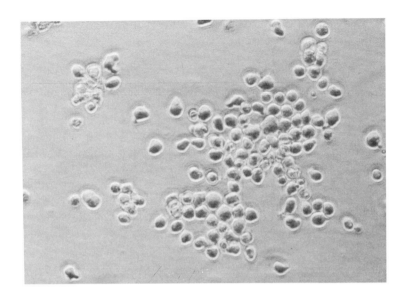

Fig. 5.1.3a Mouse myeloma cells (P3-X63-Ag8.653)

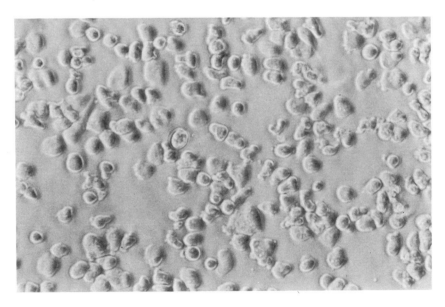

Fig. 5.1.3b Rat myeloma cells (YB2/3.0 Ag20)

from extramedullary sites of patients with advanced myeloma and have abundant rough endoplasmic reticulum (RER) and mitochondria and a well-developed Golgi apparatus associated with high immunoglobulin production and secretion. They do not carry EBV and are aneuploid. Their doubling time in culture of 36–73 hours renders the selection and growth of appropriate mutants as potential fusion partners much slower than that of murine myelomas.

As a result, the majority of human fusion partners are lymphoblastoid in origin (LCL), i.e. derived from malignant or normal haematopoietic tissue and therefore polyclonal. They are diploid, have numerous free polyribosomes, have poorly developed RER and Golgi apparatus and secrete less immunoglobulin than myelomas. The doubling time is 20–30 hours. Their relative 'success' as fusion partners may be related to a particular stage of B-cell development crucial to antibody production when fused with normal immune B-cells. The major problem with LCLs is that of sustaining Ig production. They are also invariably EBV-positive so it can be difficult to ascertain whether true hybrids or transformed lymphocytes are produced and EBV positivity would present problems in administration of the antibodies to patients.

Because of the many variables that contribute to the success or failure of establishing human hybridomas, it is difficult to compare the quality of the 'myeloma' parents directly. Specific references to individual experiments are given in Table 5.1.4, but there have been a number of comparative studies measuring parameters such as fusion efficiencies, hybrid stability and immunoglobulin secretion (Abrams et al, 1983; Cote et al, 1983; Houghton et al, 1983). More recent reviews can be found in Engelman et al (1985), Kozbor et al (1986, 1987) and James & Bell (1987). None of the lines give comparable results to the murine systems. Those that fuse with better efficiency tend to be poor Ig secretors and the better Ig secretors have poor fusion efficiencies. Only one line to date is reported to be an Ig non-producer and non-secretor but is as yet unproven as a good fusion partner (Kozbor et al, 1987).

Table 5.1.4 A selection of human fusion partners

Fusion partner	Origin	Cell type	Secreted Ig	Drug sensitivity	Reference
SKO-007	U-266	Myeloma	IgE, λ	8-AG	Olsson & Kaplan (1980)
FU-266	U-266	Myeloma	IgE, λ	8-AG	Teng et al (1983)
RPMI 8226-8AG	RPMI 8226	Myeloma	λ	8-AG	Abrams et al (1983)
RH-L4		Lymphoma	Non-secretor	8-AG	Olsson et al (1983)
GM1500-6TG-2	GM1500	LCL	IgG2, ϰ	6-TG	Croce et al (1980)
GM4672	GM1500	LCL	IgG2, ϰ	6-TG	Osband et al (1982)
GK-5	GM1500	LCL	IgG2, ϰ	6-TG	Dwyer et al (1983)
KR-4	GM1500	LCL	IgG2, ϰ	6-TG, Oua	Kozbor et al (1982)
LICR-LON-HMy2	ARH 77	LCL	IgG1, ϰ	8-AG	Edwards et al (1982)
UC729-6	W1-L2	LCL	IgM, ϰ	6-TG	Glassy et al (1983)
W1-L2-727	W1-L2	LCL	IgM, ϰ	6-TG, Oua	Emanuel et al (1984)
W1-L2-729-HF2	W1-L2	LCL	IgM, ϰ	6-TG, Oua	Strike et al (1984)
KR-12	KR4× RPMI 8226	h/h hybrid myeloma	IgG2, ϰ and λ	6-TG, Oua	Kozbor et al (1984)
SHM-D33	X63-Ag8.653 ×FU-266	m/h hetero myeloma		6-TG, Oua, G-418	Teng et al (1983)

The consensus of opinion suggests that, of the human cell lines, the best to try are LICR-LON-HMy2, UC729-6 and HF-2. Attempts to harness the stability and high fusion efficiencies of mouse myelomas by fusing the human lymphocytes did not really improve the situation. Mouse–human hybrids of this kind lose human chromosomes preferentially, thus making it difficult to stabilise hybridomas. The loss of human chromosomes is not a random process. Chromosomes 14 and 22, which code for heavy chain and lambda light chain respectively, are retained, and chromosome 2, which codes for kappa light chains, is lost (Croce et al, 1979; Erikson et al, 1981).

Better results, in terms of stable antibody production, seem to be obtainable with heteromyelomas, i.e. hybrids between human and mouse myelomas which are then fused with EBV-transformed immune lymphocytes. The lines listed in Table 5.1.4 that show the most promise are KR-12 and SHM-D33.

KR-12 is a hybrid or heteromyeloma derived from the fusion of KR-4 and RPMI 8226. RPMI 8226 is HAT resistant but KR-4 is both HAT sensitive and ouabain resistant, so hybrids are selected in medium containing HAT and 10 μM ouabain. They are then back-selected for resistance to 6-thioguanine (30 μg/ml) in a similar way to the mouse cell selection described above. Ouabain resistance was originally selected for in the KR-4 line by exposing 10^6 cells/ml to a mutagen such as ethylmethanesulphonate (60–150 μg/ml) for 24 hours or gamma-irradiation (100–300 rad). After allowing a 10-day period of expression in normal medium, they were seeded in 96-well plates at 2×10^6 cells/ml with 0.1 μM ouabain, gradually increasing to 0.5 mM over several weeks. KR-4 took five months to establish a resistance to 0.5 mM ouabain.

SHM-D33, a mouse–human heterohybridoma, was derived from the HAT-sensitive human myeloma line, FU-266, which was rendered resistant to the antibiotic G-418 by transfection with the recombinant plasmid vector, pSV2-neo^R. One of these clones was fused with the mouse myeloma line, X63-Ag8.653. Hybrid selection was effected by growth in medium containing G-418 (400 μg/ml) to kill the mouse cells and ouabain (0.5 μM) to kill the human cells. Hybrids also retain the HAT sensitivity of both parents.

5.1.5 Maintenance of myeloma cell lines

There are slight variations in culture characteristics between different myeloma cell lines. Full instructions on individual growth and maintenance conditions should be provided or obtained from the supplier of the cells. As a general rule the method described in Protocol 5.1.5.1 can be applied. Normally the cells are cultured in the presence of the antibiotics penicillin and streptomycin, which is probably the best policy in view of the long-term culture involved. However, some advocate using no antibiotics in order to encourage proper aseptic technique by the operator.

Maintaining myeloma cells in the best condition for hybridoma production not only involves their culture but also counting, viability testing and frozen storage. These procedures are described in Protocols 5.1.5.2, 5.1.5.3, 5.1.5.4 and 5.1.5.5.

PROTOCOL 5.1.5.1: MAINTENANCE OF MYELOMA CELLS

Materials:

Basic monoclonal laboratory equipment and plasticware as described in earlier
   chapters
RPMI 1640 culture medium with 10% FCS and antibiotics (Appendix A)

Method:

1. Cells are usually maintained in 250-ml culture flasks in 50 ml of complete culture
   medium with 10% FCS, in a 5% CO_2, 37°C incubator, at a density of between
   10^5 and 5×10^6 cells/ml. Too low a density results in very slow growth and too
   high a density results in overcrowding and cell death. The counting procedure
   for cells is given below.
2. When the cells have grown sufficiently (doubling time 12–48 hours), they should
   be diluted in fresh medium and distributed to additional flasks. The proportion
   of cells that grow in suspension or attach to the surface on the flasks will vary
   considerably from line to line. The mouse lines that stick can be resuspended
   easily by gentle scraping with a syringe and kwill, plastic pasteur or 'rubber
   policeman', but some rat and human lines may need to be detached by trypsinisa-
   tion (see Appendix A). The detached cells can be transferred to new flasks by
   withdrawing the cell suspension with an automatic pipette, sterile pasteur or
   20-ml syringe and kwill. Medium should never be poured from the flask.
3. It is inadvisable to maintain myeloma cells in culture indefinitely, since they
   are prone to spontaneous mutations. Thus, they may revert to their natural
   HGPRT$^+$ state. To ensure against this possibility the cells should be grown
   periodically (for a few days at monthly intervals) in medium containing
   2×10^{-5} M 6-TG or 8-AG. HGPRT$^-$ cells will be unaffected but any HGPRT$^+$
   revertants will die. Certain human myeloma lines are unstable enough to require
   continuous culture in medium containing the appropriate selection drugs.
   Obviously, these cells need thorough washing before fusion to avoid poisoning
   the hybrid cells.
4. Adequate stocks of myeloma cells should be stored in liquid nitrogen (see
   Protocol 5.1.5.4) and aliquots recovered when required. This avoids the expense
   and time involved in long-term culture and insures against infection and
   incubator breakdown.
5. Prior to fusion it is essential that the myeloma cells are in the log phase of growth.
   This can be ensured by more frequent dilution of the cells on the days leading
   up to fusion and maintenance at a cell density of not greater than 10^6 cells/ml
   and cell viability of greater than 95% (see Protocol 5.1.5.3).

PROTOCOL 5.1.5.2: COUNTING MYELOMA CELLS

Materials:

100 μl sample of cell suspension
Haemocytometer
Microscope

continued on next page

continued

The haemocytometer:

This is a heavy glass slide with two chambers, each lined with Neubauer rulings as shown in the diagram. There are nine large squares divided by triple white lines. The four corner squares are subdivided into 16 smaller squares and the central square into 25 smaller squares, each further subdivided into 16 squares. Each large square has an area of $1\,\text{mm}^2$. When a coverslip is passed over the slide so that interference patterns appear, the depth of the chamber is 0.1 mm.

The total volume over each large square is therefore:

$$1 \times 1 \times 0.1 = 0.1\,\text{mm}^3 = 0.0001\,\text{cm}^3 = 10^{-4}\,\text{ml}$$

When a suspension is applied to the chamber, the total number of cells counted in each large square represents the number in $10^{-4}\,\text{ml}$ of the suspension.

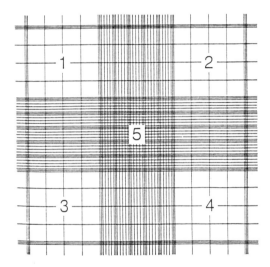

Method:

1. Resuspend the cells to be counted and apply to the haemocytometer. Normally it is unnecessary to pre-dilute myeloma cells.
2. Count all the cells in each of the four large corner squares. The total is divided by 4 to obtain the mean number in $0.1\,\text{mm}^3$ or $10^{-4}\,\text{ml}$.

   i.e. If you have 20 ml of myeloma cell suspension for fusion, an aliquot is applied to the chamber.

   Cell counts in each corner square: 53, 51, 47, 49
   Cell number in 1 ml of original suspension

   $$= \frac{53+51+47+49}{4} \times 10^4 = 50 \times 10^4 = 5 \times 10^5$$

   Total cell number $= 5 \times 10^5 \times 20\,\text{ml} = 1 \times 10^7$ cells

PROTOCOL 5.1.5.3: VIABILITY TESTS

Materials:

As for cell counting+
  ethidium bromide/acridine orange stain (see Appendix A)
  fluorescence microscope
or
  0.2% nigrosin dye (see Appendix A)

Method:

1. Take resuspended cell sample as for haemocytometer counting, and mix with an equal volume of ethidium bromide/acridine orange stain. No pre-incubation is necessary so the mixture can be applied immediately to the counting chamber.
2. Count the cells in the usual way under a fluorescence microscope. Viable cells take up acridine orange and their nuclei will fluoresce green. Ethidium bromide cannot penetrate live cells but will be taken up by dead cell nuclei which will fluoresce red.

Comments:

Both dyes are highly mutagenic and so should be handled with care. When calculating the final number of cells in the sample, remember to multiply by 2 to allow for the twofold dilution of sample with stain.

Alternative method:

(using an ordinary phase-contrast microscope)
1. Take resuspended cell sample as above and mix with an equal volume of 0.2% nigrosin dye (see Appendix A).
2. Incubate at room temperature for 5 min before counting. Viable cell membranes exclude the penetration of dye into the cell, whereas dead cells take up dye, resulting in a visual distinction.

Comments:

The commonly used dyes, trypan blue, and, to a lesser degree, eosin, have the disadvantage that they will be rapidly taken up by viable cells within 10 min of staining, to reach a plateau of about 60% viability. They also have great affinity for protein, so elimination of serum from the cell diluent would give greater accuracy. The use of nigrosin overcomes these disadvantages for most purposes, although it is not as easy for an inexperienced eye to distinguish between the brown–black of a viable stained cell and a dead cell.

PROTOCOL 5.1.5.4: STORAGE OF CELLS

A. Freezing

Desired cell density $2–10 \times 10^6$ cells in 0.25 ml freezing medium per cryotube.

Materials:

Cryotubes
Freezing medium [96% FCS, 4% dimethylsulphoxide (DMSO)]
Ice bucket
Polystyrene box
$-70°C$ freezer
Liquid nitrogen freezer

Method:

1. Resuspend cells and count sample.
2. Centrifuge at 1500 rpm for 5 min.
3. Add appropriate volume of cold freezing medium to cell pellet.
4. Resuspend and aliquot 0.25 ml to each cryotube.
5. Place in polystyrene box in $-70°C$ freezer immediately and transfer to liquid N_2 within 24 hours for long-term storage.

Comments:

Steps 2–4 should be carried out quickly to minimise DMSO contact with cells at room temperature.

Cells must be growing healthily for freezing to be successful.

Other recipes for freezing medium quote up to 10% DMSO and 90% RPMI medium with 20% FCS. Whichever is used, practice freezing myeloma cells before valuable hybridomas. The success of freezing varies according to the hybridoma line; even loss of specificity has been known to occur.

B. Thawing

Materials:

56°C waterbath
Centrifuge tubes containing approx. 20 ml cold RPMI medium without serum
24-well culture plate
Thymocyte medium

Method:

1. Thaw cryotube as quickly as possible in 56°C waterbath.
2. Dilute cells gradually with cold medium (20 ml).

continued on next page

continued

3. Centrifuge at 1500 rpm for 5 min.
4. Resuspend pellet in 1 ml warm thymocyte medium and transfer to one well of 24-well culture plate.
5. Inspect cells after settling and expand if necessary.

Comment:

Thymocyte medium is not always necessary, especially with myeloma lines and well-established hybridomas, but it is better to be safe than sorry.

PROTOCOL 5.1.5.5: FREEZING HYBRIDOMAS DIRECTLY IN MICROTITRE PLATES (after Wells & Price, 1983)

This provides a simple and convenient means of preserving emerging clones at the earliest possible stage, and allows the operator more control over the numbers of cell lines to be maintained in culture whilst preliminary characterisation is carried out.

To freeze:

Materials:

96-well plate containing growing macroscopically visible cells
Freezing medium (culture medium + additives + 6% DMSO)
Plastic 'Clingfilm'
Insulated Jiffy bag

Method:

1. Feed cells with fresh medium the day before.
2. Aspirate growth medium and replace with 50 μl freezing medium per well.
3. Wrap the plate in 'Clingfilm' and place inside Jiffy bag.
4. Freeze immediately by transfer to −70°C freezer.

To thaw:

Materials:

RPMI wash medium
RPMI culture medium
Feeder cells

Method:

1. Add 150 μl of prewarmed (37°C) RPMI wash medium to each well and incubate in CO_2 incubator for 5 min.
2. Aspirate medium and refeed wells with 200 μl of culture medium with feeder cells (10^4 per well).
3. Feed and propagate as usual according to cell growth.

5.2 LYMPHOCYTES

In rodent fusions, the preferred lymphocyte donor is the spleen, in which approximately 20% of cells are B-lymphocytes. It is estimated that no more than 10^5 of these are likely to secrete specific antibody. Nevertheless, obtaining enough specific hybridomas from this pool is not usually a problem despite low fusion frequencies ($1 : 10^6$). However, there are occasions when enriching the specific B-lymphocyte population prior to fusion would be useful. For instance, pooled B-lymphocytes from several weakly immune spleens could be fused together in one operation. Enrichment strategies, of which a variety are described below, are more widely applied to lymphocytes from human peripheral blood, where activated splenocytes are rarely available.

5.2.1 Spleen cells

There are several commonly used methods for preparing cell suspensions from organs, including the use of teasing forceps and pressing tissue through a fine sieve. The method described here (Protocol 5.2.1), of dispersion by fine jets of fluid, may take a little longer (10 min) but in the authors' experience yields a better single cell suspension. One mouse spleen yields about 5×10^7 to 2×10^8 nucleated cells, whereas a rat spleen may be expected to yield approximately three times as many cells.

PROTOCOL 5.2.1: PREPARATION OF A SPLEEN CELL SUSPENSION

Materials:

Balb/c mouse
Ether jar
70% ethanol
Sterile fine curved forceps
Sterile scissors
Universal container with approx. 5 ml sterile saline
Small Petri dish (35 mm)
Scalpel
2×10-ml syringes containing 10 ml each culture medium ($-$FCS)
2×25-g needles
50-ml sterile centrifuge tubes

Method:

1. Kill the mouse by cervical dislocation or ether anaesthesia.
2. Swab the mouse liberally with 70% ethanol and place on its right side.
3. Using sterile scissors and forceps, cut the skin in the inguinal region, exposing the peritoneal wall.
4. Cut through the peritoneal wall, exposing the spleen.
5. Lift the spleen with forceps and release from surrounding tissue.
6. Place the spleen in the container of sterile saline.
7. Transfer the spleen to a small Petri dish in a laminar flow hood.
8. Make a superficial cut with the scalpel along the ridge of the spleen.

continued on next page

---

continued

9. Two syringes fitted with 25-g (fine) needles are then used to inject culture medium (10 ml each) vigorously at multiple sites into the spleen such that the spleen cells are seen to become dispersed. After this procedure, although the spleen will retain its shape, it will be noticeably paler. Further disruption of the spleen is unnecessary to achieve the required yield.
10. The medium containing dispersed cells is transferred to a sterile centrifuge tube, any large aggregates and tissue pieces being discarded. The cells are centrifuged at 400 g (i.e. 1500 rpm on a bench centrifuge) for 10 min.
11. The supernatant is removed and discarded. The cells are then resuspended in 10 ml of wash medium and the washed cells collected by centrifugation as above.
12. The washed cells are resuspended and used immediately for fusion.

5.2.2 Peripheral lymphocytes

The simplest way of isolating lymphocytes from peripheral blood is by density centrifugation using a Ficoll/sodium metrizoate medium such as Ficoll-Hypaque (Pharmacia) or Isopaque (Nyegaard) as developed by Boyum (1968). Ficoll-Hypaque is an aqueous solution of a synthetic high molecular weight polymer of sucrose and epichlorohydrin mixed with sodium diatrizoate and EDTA. It has a density of 1.077 ± 0.001 g/ml optimised for the most efficient yield of human mononuclear cells, i.e. $60 \pm 20\%$ of the original blood sample. Defibrinated or anticoagulated blood is diluted appropriately in a balanced salt solution and carefully layered onto the Ficoll-Hypaque in a centrifuge tube (see detailed protocol). After a short centrifugation period, lymphocytes, monocytes and platelets remain at the interface. Subsequent washing removes the platelets and residual Ficoll. Monocytes can be removed if required by incubating the blood sample with iron or iron carbonyl particles before separation. The monocytes become denser after phagocytosing the particles and are thus, on centrifugation, sedimented to the bottom of the tube with the red blood cells and granulocytes. T-cells can also be removed prior to centrifugation by using the SRBC rosetting method (see section 5.2.3.1).

PROTOCOL 5.2.2.1: LYMPHOCYTE SEPARATION FROM HUMAN PERIPHERAL BLOOD

The quality of separation is dependent mainly on the speed of centrifugation and the height of the blood sample in the tube. This protocol deals with a blood volume per tube of 2 ml (4 ml when diluted) which should yield 10^6 lymphocytes/ml of which 20% should be B-cells. If a mouse spleen is used (10^8 cells) approximately 50% will be B-cells. Varying the protocol for larger volumes will require a greater appreciation of the technique than can be covered here. Readers are referred to the manufacturers' instructions.

continued on next page

continued

Materials:

Fresh venous blood anticoagulated (i.e. EDTA, see Appendix A)
Balanced salt solution (see Appendix A)
Ficoll-Hypaque
15-ml tissue culture sterile plastic centrifuge tubes
Pasteur pipettes (glass and sterile plastic)
Complete RPMI medium

Method:

1. Dispense 3 ml of well-mixed Ficoll-Hypaque at 18–20°C into the centrifuge tube.
2. Dilute 2 ml of the blood (at 18–20°C) with an equal volume of balanced salt solution and layer it gently onto the surface of the Ficoll. *Do not mix.*
3. Centrifuge at 400 *g* for 30–40 min at 18–20°C.
4. Draw off the upper layer with a clean pasteur pipette, leaving the lymphocyte layer at the interface undisturbed.
5. Using a clean sterile plastic pasteur pipette, remove the lymphocyte layer in a minimum volume and wash twice in at least three volumes of balanced salt solution with centrifugation at 60–100 *g* for 10 min.
6. Resuspend the cells in complete RPMI.

Comment:

Although Ficoll-Hypaque is designed for use with human lymphocytes it can also be used to separate lymphocytes from mouse and rat. However, better yields can be obtained by using Percoll (Pharmacia) in which the density of the medium can be tailored to optimal yields (a density of 1.08 is usually made up for mouse cells) (Chi & Harris, 1978; Mizobe et al, 1982).

5.2.3 Lymphocyte selection

5.2.3.1 T-CELL DEPLETION (ROSETTING METHOD)

Human T-lymphocytes can be distinguished from B-lymphocytes by their ability to form spontaneous rosettes with sheep red blood cells. This property, together with density centrifugation, has been used extensively as a non-specific means of separating these two groups of human lymphocytes. Mouse (and rabbit) T-cells do not bind SRBCs spontaneously. In the context of hybridoma production, separating T- and B-lymphocytes is desirable (see Protocol 5.2.3.1), if it is intended to transform B-cells with EBV, since T-cells are often toxic to the virus. However, for in vitro immunisation and conventional fusions, separation can be detrimental if helper T-cells are removed.

PROTOCOL 5.2.3.1: T-CELL DEPLETION

In practice, to give more reproducible results, SRBCs are usually pretreated with enzymes or sulphydryl reagents. The following method is based on that described by Kaplan & Clark (1974).

Materials:

10-ml 2-aminoethylisothiouronium bromide (AET)
    10.2 mg/ml in distilled H_2O, adjust to pH 9 with NaOH, filter-sterilise
2 ml fresh-packed SRBCs. Washed three times in NaCl
0.15 M NaCl
Serum-free RPMI medium
Sterile tissue culture plastic centrifuge tubes
Ice bath

Method:

1. Mix 2 ml SRBCs and 8 ml AET solution for 15 min at 37°C and wash four times by centrifugation at 300 g for 10 min in serum-free RPMI medium. Resuspend as 2% solution (approximately) 4×10^8/ml.
2. Wash lymphocytes in RPMI medium, to remove leached surface molecules that might interfere with the reaction. Adjust concentration to 5×10^6/ml.
3. Add equal volume of SRBCs to lymphocytes at a ratio of 50 or 100 : 1, SRBCs to lymphocytes, followed by equal volumes of RPMI medium and FCS. Mix gently but thoroughly and centrifuge at 200 g for 10 min. Leave tube to stand in ice for 90 min.
4. About 60% of lymphocytes should have formed rosettes, which can be separated from B-lymphocytes on Ficoll-Hypaque as described above.

5.2.3.2 POSITIVE SELECTION OF SPECIFIC B-CELLS

B-lymphocytes with specific antibodies on their cell surface can be selected positively from other B-lymphocytes by incubation with SRBCs coated with specific soluble antigen (Protocol 5.2.3.2). Antigen coating is normally done with tannic acid (Boyden, 1951), chromic chloride (Goding, 1976), carbodiimides (Johnson et al, 1966) or glutaraldehyde covalent coupling techniques (Walker et al, 1979). The specific B-cell rosettes are then separated by density centrifugation as described in Protocol 5.2.2.1, but, for use in fusion, the SRBCs need to be removed. If the SRBCs are coated but not tanned they can be lysed by immersing the cells in distilled water for five seconds followed immediately by complete culture medium.

**PROTOCOL 5.2.3.2: POSITIVE SELECTION OF SPECIFIC
B-CELLS USING ANTIGEN-COATED SRBCs**

The optimal conditions for a particular protein should be determined
experimentally.

Materials:

  SRBCs in Alsever's solution (up to two weeks old)
  0.15 M NaCl
  Antigen solution (1 mg/ml of soluble protein)
  1% (w/v) matured solution of hydrated chromic chloride in 0.15 M NaCl, pH 5.
    Keep for three weeks, readjusting pH. Stock stable at room temperature for
    two years. Dilute 1 : 100 before use.
  PBS

Method:

1. Wash SRBCs by centrifugation at 400 *g* for 10 min in 0.15 M NaCl, at least three
   times or until the supernatant loses its red colour. Remove buffy coat above
   cell pellet, to minimise spontaneous rosette formation between SRBCs.
2. Dilute 0.5 ml of packed cell volume to 4.5 ml with 0.15 M NaCl.
3. Add 0.5 ml of antigen solution.
4. To this mixture, add 5 ml dropwise of diluted chromic chloride solution with
   constant vortex mixing for 15 min.
5. Add 10 ml PBS (phosphate inhibits the reaction), mix and centrifuge at 400 *g*
   for 10 min. Repeat twice.
6. Store at 4°C before use.

5.2.3.3 NEGATIVE SELECTION OF SPECIFIC B-CELLS

A similar principle can be used to deplete a cell mixture of unwanted B-cells, especially
when the specific antigen is cellular or ill-defined, so that positive selection cannot
be used. B-cells, when in the presence of saturating quantities of antigen ($100 \, \mu g/10^7$
cells/ml), will lose their surface immunoglobulin, a phenomenon known as 'capping'.
Subsequently, if SRBCs coated with $F(ab)_2$ fragments of anti-species immunoglobulin
is added to the mixture, the SRBCs should only bind to the non-specific B-cells, which
can be removed by density centrifugation (Walker et al, 1977). It should be remembered,
though, that the ability of a cell to bind antigen is not necessarily correlated with its
ability to secrete antibody (Nossal & Pike, 1976; Kozbor & Roder, 1981).

All the applications described above using coated SRBCs can also be adapted for use
with magnetic or fluorescent beads, depending on the hardware available for separation
(Owen et al, 1979; Owen, 1981; Dangl & Herzenberg, 1982). These beads are also
obtainable ready-coated with monoclonal antibodies specific for lymphocyte subsets,
if such separation is desired (Dynal).

5.2.4 Epstein–Barr virus transformation

As an alternative to the traditional means of monoclonal antibody production, there is a completely different method of immortalising human cells, which has received more attention as a result of the problems with human fusion partners. This involves activation and transformation of lymphocytes by Epstein–Barr virus (EBV) (Steinitz et al, 1977; Roder et al, 1986; Walls et al, 1988). EBV is a human B-lymphocytic herpes virus which is carried by an estimated 90% of the adult population. In vitro, it is capable of stimulating the growth of B-lymphocytes (transformation or immortalisation), thus perpetuating the properties of those lymphocytes, including immunoglobulin secretion. It is possible to obtain specific antibody-producing cultures by infecting lymphocytes from recently immunised individuals with EBV. Antibody levels can reach 20–50 μg/ml, mainly IgM. However, there can be problems, due to the presence of memory T-lymphocytes from EBV-infected individuals, which are cytotoxic when stimulated in vitro. If the lymphocytes are seeded at high densities of $1-2 \times 10^6$/ml, the proliferating foci of B-lymphocytes regress after three to four weeks. Regression can be avoided

---

PROTOCOL 5.2.4: EBV TRANSFORMATION OF LYMPHOCYTES

The usual source of EBV is the cell line B95-8 (Miller et al, 1972), that was derived from marmoset peripheral blood mononuclear cells infected in vitro with wild-type virus obtained from a patient with infectious mononucleosis. B95-8, obtainable from ECACC or ATCC (see Appendix B), is grown to confluence in RPMI medium containing 10% FCS and the culture supernatant harvested. Stocks of supernatant can be stored at $-70°$C until required. If required, the viral activity of the supernatant can be tested by incubation with BJAB cells and measurement of EBV-specific nuclear antigen (EBNA) expression (Kozbor et al, 1986).

Materials:

> Target B-cells (harvested and enriched by rosetting and Ficoll separation as described above)
> EBV (1 ml undiluted, 0.45 μm filtered B95-8 supernatant or 10^7 transforming units per 10^6 lymphocytes)
> RPMI 1640 culture medium, 10% FCS
> Sterile centrifuge tubes
> Feeder cells
> 96-well plate

Method:

1. Incubate appropriate volume of EBV supernatant with 0.5 ml of sedimented lymphocytes for one hour with frequent suspension.
2. Centrifuge the cells, wash and resuspend in fresh culture medium at a density of 10^4/ml or 10^3 cells per well of a 96-well plate.
3. Add feeder cells such as peripheral blood mononuclear cells from a EBV-negative donor (10^4 per well).
4. Culture under the usual conditions until screening and cloning.

by seeding at lower densities ($1–5 \times 10^5$/ml) or by eliminating the T-lymphocytes by rosette depletion, inhibition by cyclosporin or adding phytohaemagglutinin to the medium at the start to inhibit specific T-cell activation. If T-cells in general are depleted, feeder cells need to be added to the culture to support immortalisation of the B-cells (Pope et al, 1974), such as autologous or allogeneic peripheral blood monocytes or foetal fibroblasts, X-irradiated prior to culture (2000 R).

Even with these precautions, more stable antibody production has been obtained in recent years by applying EBV transformation in combination with conventional fusions, either with mouse or human partners or with mouse–human heteromyelomas (Teng et al, 1983).

5.3   CELL FUSION

Before considering the technical problems associated with the production of antibody-secreting hybrids, it is appropriate to look at some general aspects of cell hybridisation. The generation of hybrid cells was already possible before Kohler and Milstein applied the technique. In fact the technique is pivotal to the whole field of genetic analysis in cultured animal cells. It was discovered originally that under particular conditions two somatic cells can be caused to fuse in vitro to produce a heterokaryocyte, which is a single hybrid cell containing two or more nuclei. What is particularly relevant to both genetic analysis and hybridoma technology is that a proportion of heterokaryocytes will multiply indefinitely. The first mitotic division after fusion will lead to a population of daughter cells that carry both sets of parent cell chromosomes in the same nucleus. Often, however, at subsequent divisions chromosomes are lost in ones and twos until a new stable cell line is formed carrying some chromosomes from each parent.

In order to capitalize upon this phenomenon the first technical problem to be overcome is how to increase the otherwise very low frequency of spontaneous cell fusions. For the generation of hybridomas the original agent used to enhance the frequency of fusions was inactivated Sendai virus (Kohler & Milstein, 1976). This phenomenon is based on the observation that when cells to which virus is attached are incubated at 37°C the cell membrane adjacent to the site of virus adsorption becomes damaged and cytoplasmic bridges develop between adjacent cells. As these sites of damage are repaired the bridges enlarge and eventually fused cells form. This technique has disadvantages, including its ineffectiveness with some types of cell, the laborious nature of virus production, the variability between batches and the possibility of introducing viral genes into the cells. There is also a possibility that the system is biased towards IgM-producing hybrids.

Various alternative chemical fusing agents for mammalian cells have been sought in the past. Simple chemicals such as lysolecithin and polyethylene glycol (PEG) were found to be effective (Pontecorvo, 1975) and currently PEG enjoys almost universal use.

5.3.1   Polyethylene glycol-aided fusion procedures

A number of different protocols for the use of PEG in hybridoma fusions have been described (Galfre et al, 1977; Gefter et al, 1977; Oi et al, 1978) and the differences

PROTOCOL 5.3.1: POLYETHYLENE GLYCOL-AIDED FUSION
(after Galfre & Milstein, 1981)

Materials:

10^7–10^8 myeloma cells in log phase
10^8 immune spleen cells
RPMI wash medium (50 ml)
PEG 1500 40% 0.8 ml (Appendix A)
RPMI culture medium (20% FCS) 30 ml
2×20-ml syringes
2×25-g needles
Kwills
50-ml centrifuge tubes
3×96-well culture plates

Method:

1. Centrifuge both cell populations at 1500 rpm for 5 min.
2. Resuspend each in 10 ml serum-free RPMI medium and mix together.
3. Centrifuge at 1500 rpm for 5 min.
4. Remove supernatants completely (invert tube and tap out excess fluid).
5. Add 0.8 ml 40% PEG 1500 via a 25-g needle, slowly over 1 min, and stand for 1 min.
6. Add 1 ml wash medium (via 25-g needle) slowly over 1 min, followed by 20 ml over 5 min.
7. Centrifuge at 1500 rpm for 15 min.
8. Discard supernatant and resuspend gently in 15 ml complete RPMI culture medium.
9. Distribute cell mixture between 3×96-well plates (180 wells). The outside wells of the plates are filled with wash medium.
10. Top up medium in each well with more culture medium (15 ml total), i.e. $100 \, \mu l$ per well.
11. Leave plates in 5% CO_2, 37°C incubator overnight before adding HAT medium.

Comments:

1. The FCS must be pretested for its ability to support clonal growth (see Protocol 6.2).
2. After addition of PEG all handling of the cells should be as gentle as possible to minimise dispersal of forming hybrids.

between the various approaches have been compared in detail by Fazekas de St Groth and Scheidegger (1980).

The variables considered most important include the concentration of PEG employed, the pH of the mixture and the length of time of exposure. The concentrations of PEG used are normally between 40% and 50% (w/v) because the frequency of fusion below 30% is minimal whereas above 50% toxicity becomes a problem. Maximal fusion frequencies occur between pH 8 and pH 8.2 but because of the toxic nature of PEG the time of exposure to the reagent is kept to a minimum (i.e. 1–2 min in 50% PEG; up to 7 min in 30% PEG). In addition, the temperature must remain between 20°C and 37°C for successful fusion, the ratio of cell types used being between 1 : 1 and 10 : 1 lymphoid to myeloma cells. Normally PEG of molecular weight 1500–4000 is used.

5.3.2 Electrofusion

In recent years a new method of mammalian cell fusion has been developed (Zimmermann, 1982, 1986). The electrofusion technique, in which cells are aligned by dielectrophoresis, has been applied to hybridoma production using both mouse and human cells (Bischoff et al, 1982; Ohnishi et al, 1987; Foung & Perkins, 1989; Borrebaeck & Hagen, 1989). In this technique, cells are drawn together by dielectrophoretic effects in a non-uniform alternating electric field of low field strength. Fusion of apposed cells occurs when the cell suspension is exposed to a high, rectangular electric pulse of perhaps 10 μs duration. The pulse causes dielectric breakdown of the cell membranes and also increases the intimate cell contact already achieved by dielectrophoresis. However, a number of disadvantages are associated with this procedure. Firstly, an unphysiological cell-suspending phase of low ionic strength is required, in order to avoid undesirable temperature increases. Secondly, cell chains produced by dielectrophoresis tend to migrate towards the electrodes. This means not only that passage of the fusion pulse through cell chains may give rise to a fusion product derived from more than two single cells, but also that crowding of cells at the electrodes may place a limit on the number of cells which may be treated and, consequently, on the interelectrode distance. The practical upper limit on this distance (22 μm) reduces the volume of cell suspension that can be treated to between 10 and 200 μl. Even at the upper limit of volume, the total number of cells is confined to 10^6 (Vienken & Zimmermann, 1985). This latter compares unfavourably with PEG-induced fusion mixtures in which, generally, 3×10^8 splenocytes are mixed with between 3×10^7 and 3×10^8 myeloma cells.

Clearly an improved or alternative fusion method using more practical cell numbers and physiological solutions, and which is applicable to cells from all the relevant species, is required. One method, with significant potential, that would appear to meet these conditions, is electrofusion with cell alignment based on ultrasonic sound fields (Vienken et al, 1985; Vienken and Zimmermann, 1985).

5.3.3 Electroacoustic fusion

Cell aggregation in a megahertz sound field arises from the interaction of the effects of three radiation forces. One field causes cells to be attracted to each other. A second field exerts a torque on the cells which tends to align them in a preferred direction. A third operates in an acoustic standing wave, causing the cells to concentrate at positions separated by half a wavelength. Since these radiation forces show different dependencies on acoustic pressure and on frequency (Nyborg, 1977), these parameters can be varied systematically in order to achieve maximum cell alignment for different cell types.

A system has been developed in which cells can be levitated on a 1-MHz standing wave sound field (Coakley et al, 1989; Bardsley et al, 1989, 1990). Application of the same technology to cells in a concentric cylinder system of 200-μm gap size (designed for dielectrophoresis fusion) has given high fusion yields with pronase-treated erythrocytes and with myeloma cells (Vienken et al, 1985; Vienken and Zimmermann, 1985). It was noted (Vienken et al, 1985) that many of the fused cells arose from doublets and triplets which remain in suspension away from the container walls with little evidence of the multiple cell fusion and migration which can occur with dielectrophoresis.

These factors and others remove the limitation of the short interelectrode gap which is associated with the dielectrophoresis method. Consequently, large volumes of cell suspension containing large numbers of cells may be exposed to the fusion pulse in the electroacoustic system. The increase in volume and the larger interelectrode gap necessitate the use of a 10-kV fusion generator (rather than the 250-V pulse available in the commercial Zimmermann electrofusion system). Such high-voltage pulse generators have already been used successfully by Berg (1982) to produce yeast hybrids at much enhanced efficiencies, in a 1-ml volume, by electrofusion in the presence of PEG.

5.3.4 Fusion enhancement strategies

Lo et al (1984) published a method whereby specific antibody-secreting lymphocytes could be brought together with myeloma cells, prior to electrofusion, thus increasing the chances of specific antibody-secreting hybrids being formed. This was achieved by covalently binding biotin to the myeloma cell surface and binding antigen–avidin complexes to the specific antigen receptors expressed on the B-cell surface. The biotin–avidin interaction thus brings the cells together. A technically simpler version of this idea was put forward by Wojchowski and Sytkowski (1986) (see Fig. 5.3.4), in which biotinylated antigen was bound to B-cells, incubated with streptavidin and then brought together with biotinylated myeloma cells, before electrofusion.

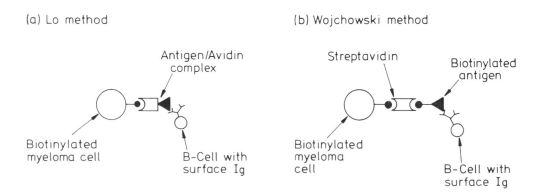

Fig. 5.3.4 Fusion enhancement strategies

Chapter 6

Hybrid selection and cloning

6.1 HYBRID SELECTION

The cells of the fusion mixture are grown in a selective medium (i.e. HAT) such that, in time, only myeloma–lymphocyte hybrids will survive (see Protocol 6.1.1). Within approximately five days following fusion, small clusters of cells resembling myeloma cells are likely to be hybridomas. Within 10–14 days, these will have grown sufficiently

Fig. 6.1a Cell mixture immediately after fusion. Magnification ×472

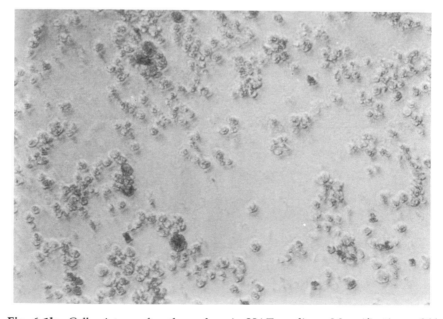

Fig. 6.1b Cell mixture after three days in HAT medium. Magnification ×944

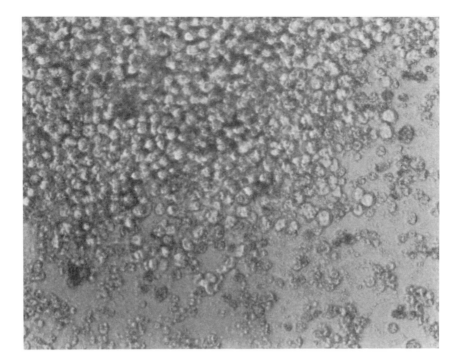

Fig. 6.1c Cell mixture after seven days in HAT medium. Magnification ×472.

PROTOCOL 6.1.1: HAT SELECTION

Materials:

HAT medium (Appendix A)
Fusion plates (from Protocol 5.3.1)

Method:

1. The 96-well plates are changed to HAT medium on the day following fusion (day 1). Half the old medium is removed (i.e. with a multipipetter) and replaced with HAT (syringe and needle or multidrop pipette).
2. This medium change is repeated on day 2, 4, 6, 8, 11 and every two to three days afterwards until the first screen.
3. Within three to four days all the myeloma cells should be dead. From about the fifth day, hybrids should start to become visible.
4. Screen for specific antibody production (eleventh to fourteenth day). Allow at least three days between screen and last medium change.
5. After screen, select positive, antibody-secreting wells and expand (see below).

Testing of HAT medium:

1. Grow myeloma cells (from same flasks as those in fusion) in HAT medium. They should all be dead in four days.
2. Grow hybrid cells in HAT medium. Survival indicates HT OK.

to have secreted enough detectable antibody into the culture supernatant. Those secreting specific antibody can be identified by application of the screening test developed as described in Chapter 4. A manageable number of wells is then expanded (allowed to grow) prior to cloning and/or frozen storage. Protocol 6.1.2 illustrates how to expand cultures from the small microtitre wells to larger wells.

PROTOCOL 6.1.2: EXPANSION OF CULTURE PRIOR TO CLONING

After the first screen reveals wells containing specific antibody-secreting cells, a manageable number of selected cultures should be expanded into larger wells, so that there are enough cells to give a reliable count prior to cloning. Simultaneously, HAT medium is replaced with HT medium for several passages before returning the cultures to the usual complete medium, in order to ensure that no residual aminopterin remains in the wells.

Materials:

  Fusion plate
  24-well culture plate
  HT medium (Appendix A)
  Feeder cell suspension (i.e. one thymus in 1 ml HT medium)
  Sterile plastic Pasteur pipette (wide bore)

Method:

1. Add 1 ml HT medium to each 2-ml well required.
2. Add a few drops of thymocyte suspension (10^7 thymocytes per well) to each well.
3. Resuspend cells in positive 200-μl well (using sterile plastic Pasteur pipette).
4. Transfer to 2-ml well and mix.
5. Return 200 μl of this cell mixture to the original well.
6. Repeat with each positive well.
7. After two days feed with an additional 0.5 ml of HT medium.
8. Two days later remove as much supernatant as possible and add fresh HT medium.
9. When cells are nearly confluent (up to about one week) reassay new wells and the original wells.
10. If they are still secreting specific antibody, take sample to clone.
11. Expand remaining cultures to freeze as soon as possible (i.e. 2×confluent 2-ml wells pooled in one cryotube).

Comment:

The timing of stages 7–9 will vary considerably depending on the number of cells expanded. A feeder cell suspension or equivalent growth factor may not always be necessary, particularly when expanding large clones, but their use will improve the chances of success.

6.2 CLONING

The purpose of cloning is to isolate a single specific antibody-secreting cell from the hundreds of other cells in the culture so that a monoclonal cell line can be established. Three basic cloning techniques may be identified, but the need to freeze and store samples of all the cultures prior to and during cloning must not be overlooked. This requirement for continued storage of cells at all stages is essential for the following reasons. Firstly, throughout cloning, particularly during the early stages, there is a high probability of chromosome loss and consequent loss of immunoglobulin secretion from the hybrids. Secondly, there is always a risk of overgrowth by non-producing hybrids. Thirdly, the risk of infection and/or mechanical breakdowns should be protected against. The general procedure for the storage of cells is described later.

Before describing the cloning procedures in detail a prerequisite in all cases is the need for a medium in which hybridoma cells may be grown from very low (clonal) densities. Usually the growth of hybrids is inhibited at high cell dilutions, but this can be overcome by the use of feeder cells such as thymocytes, peritoneal macrophages or spleen cells. Thus hybrids are seeded into culture wells containing such cells.

PROTOCOL 6.2: TEST OF FOETAL CALF SERUM FOR ABILITY TO SUPPORT HYBRID CLONES

Materials:

  RPMI wash medium
  Thymocytes (one thymus per 96-well plate)
  96-well microtitre plate
  RPMI 1640+10% test or control (negative and positive) FCS (5 ml each)
  Hybrid cells ($\sim$ 120 per FCS sample)

Method:

1. Prepare thymocyte medium using RPMI medium without FCS; 1 ml for each sample of FCS.
2. Add enough hybrid cells to give 10 cells per 100 μl per well.
3. Add 100 μl of this mixture to each well (12 wells or one row per FCS sample).
4. Prepare 5 ml each of RPMI medium+10% test FCS including two controls (negative and positive).
5. Add 2 ml of these media to each row (final concentration FCS=5%).
6. After five days, score each well for hybrid growth. There should be sufficient growth for a decision to be made on the batch.
7. If not, feed the cultures on the fifth day and score later.

Comment:

The final concentration of FCS is lower than the usual working concentration but it will be easier to distinguish a good FCS from others.

In addition the quality of FCS used as a medium supplement is of critical importance. Each new batch of FCS must be tested for the ability to support hybrid clones (see Protocol 6.2). Only 10% of batches are suitable. Although several companies pretest their serum for hybridoma growth, there can be considerable variation in quality (and cost). Try samples from each before buying. Normally a minimum of five litres will be reserved for you.

6.2.1 Feeder cells and growth factors

Feeder cells are necessary to encourage the growth of hybrids during cloning, expansion from microwells and thawing from frozen storage (Coffino et al, 1972). Their precise role is still not well understood. The preparation of two commonly used types of feeder cells is illustrated in Protocols 6.2.1.1 and 6.2.1.2. Cell–cell interaction was thought to play an important role (Lernhardt et al, 1978) but enriched (conditioned) medium from thymocyte cultures (Reading, 1982) and human endothelial cell culture supernatant (Astaldi et al, 1980) have also been used successfully to improve hybridoma growth, suggesting that the effect is dependent on the production of soluble factors.

A soluble hybridoma growth factor (HGF) found in conditioned media has recently been identified (Aarden et al, 1985; Van Snick et al, 1986; Bazin & Lemieux, 1987) which is biochemically similar to a plasmacytoma growth factor, B-cell stimulatory factor 2 (BSF-2) and interferon-β_2. Poupart et al (1987) propose that it should be called interleukin-6 (IL-6). The effects of IL-6 are influenced by different batches of foetal calf serum but evidence suggests that the proportion of antibody-secreting hybridomas is increased in the early stages of culture rather than the overall number (Bazin & Lemieux, 1989). It is possible that the expression of immunoglobulin genes in these cells may be induced or stabilised in a way analogous to the role of BSF-2 on EBV-transformed B-lymphocytes. Recombinant IL-6 can now be obtained commercially (i.e. ICN).

The most commonly used feeder cells are thymocytes, splenocytes or peritoneal macrophages, the preparation of which is described in the accompanying protocols. Thymocytes, of course, can only be obtained from young animals although it is not necessary to use the same species as the hybridoma parents.

PROTOCOL 6.2.1.1: PREPARATION OF THYMOCYTE FEEDER CELLS

Materials:

3–10-week-old mouse
Sterile fine curved forceps × 2
Sterile scissors
70% ethanol
Ether jar
Universal container with approximately 5 ml sterile saline
Petri dish
Teasing forceps
15-ml sterile centrifuge tubes
Serum-free culture medium

Method:

1. Kill the mouse by ether anaesthesia.
2. Swab the mouse liberally with 70% ethanol and place on its back.
3. Using sterile scissors and forceps, cut the skin along the mid-line of the thorax up to the neck.
4. Cut through the sternum and diaphragm.
5. The two thymus lobes are whitish organs lying just above and either side of the heart.
6. Lift each lobe with a forceps and pull it out using a curved forceps underneath the lobe.
7. Place in container of sterile saline or medium and transfer to laminar flow hood.
8. Place lobes in Petri dish and remove any blood clots.
9. Tease apart and collect cell-containing medium.
10. Centrifuge at 1500 rpm (400 g) for 5 min. Discard supernatant and resuspend pellet in appropriate volume of complete culture medium.

Comments:

1. In this age range, $1–2 \times 10^8$ thymocytes per mouse should be obtained.
2. For expanding hybrids use 10^7 thymocytes per 2-ml well, i.e. approximately 1 ml of medium per thymus, and add two drops to each well.
3. For cloning use 10^7 thymocytes per ml, i.e. 20 ml of medium per thymus. Use 10 ml for each clone plate.
4. To minimise contamination of the preparation with red blood cells the mice could be heavily bled by cardiac puncture before dissection.
5. If necessary, parathymic nodes can be identified and removed by injecting the mouse with 0.1 ml of indian ink solution (1 : 5 in saline) i.p., 20–30 min before removing the thymus. The parathymic nodes will become black.

PROTOCOL 6.2.1.2: PREPARATION OF PERITONEAL MACROPHAGES FOR USE AS FEEDER CELLS

Materials:

Adult mice (not necessarily Balb/c), $\sim 10^6$ cells per mouse
Cold RPMI medium ($-$FCS) containing 10 U/ml heparin, 2 ml per mouse
Small (3-ml) plastic tube with perforations at base
Sterile plastic Pasteur pipette
15-ml sterile centrifuge tube
Ice box
70% ethanol

Method:

1. Kill mouse by cervical dislocation and swab thoroughly with 70% ethanol.
2. Make small incision in lower abdomen.
3. Insert perforated tube.
4. Add 2 ml RPMI medium, without spillage.
5. Gently massage abdomen for approximately 2 min.
6. Remove fluid using Pasteur pipette inside tube (at least 1.75 ml).
7. Pool fluid from each mouse to obtain required number of cells.
8. Spin at 1500 rpm (400 g) for 1 min.
9. Resuspend in complete RPMI medium (i.e. $+$FCS at 37°C) and use as required.

For expanding, $\sim 200\,000$ cells in 0.5 ml medium per 2-ml well
For cloning, $\sim 100\,000$ cells per well of 96-well plate

6.2.2 Cloning by limiting dilution

This is the simplest and most popular method of hybridoma cloning (Protocol 6.2.2). In practice, there are a number of different procedures but the ultimate objective is to set up at least three very low dilutions of cells resulting in a mean of one cell per well. Within two weeks of cloning, clones should be visible macroscopically and, subjectively at least, some of the wells may be considered to contain monoclonal growth. At this stage each of the cell supernatants should be screened to determine which of them contains the desired antibody. Positive wells which are likely to be of monoclonal origin are expanded, as described elsewhere in this manual. They are then recloned, with a proportion of the cells being frozen and stored. The recloning is repeated at least three times, after which the final clone plate should display positive antibody production in virtually all the seeded wells. Cells in the first clone plate producing the desired antibody even though they are unlikely to be of monoclonal origin should not be discarded. The contents of such wells may also be expanded and the cells stored so that they may be cloned at a later stage, should the necessity arise.

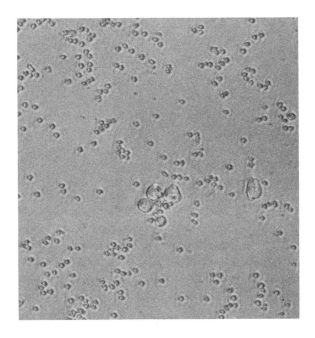

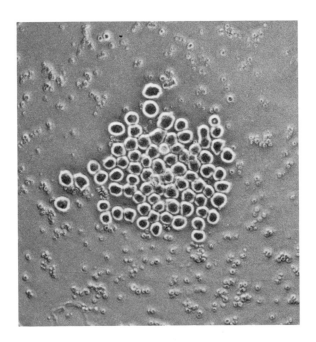

Fig. 6.2.2b Clone at seven days. Magnification ×944

PROTOCOL 6.2.2: CLONING BY LIMITING DILUTION

Materials:

  Cells to be cloned
  96-well plate
  RPMI wash medium
  RPMI culture medium (20% FCS)
  Thymocyte medium (10^7 per ml)
  (10 ml growth medium + 1 ml thymus lobe per plate)
  15-ml sterile centrifuge tubes

Objective:

  Plate 24 wells with average 5 cells per well
  Plate 24 wells with average 2 cells per well
  Plate 24 wells with average 1 cell per well
  Plate 24 wells with average 0.5 cells per well

Method:

1. Resuspend and count sample of cells to be cloned and simultaneously remove 100-μl sample of same suspension to 15-ml centrifuge tube for cloning.
2. Dilute cloning sample in RPMI (−FCS) to 2000 cells/ml.
3. Take a 100-μl, well-suspended sample of this and add to 4 ml of thymocyte medium, i.e. a total of 200 cells.
4. Plate 24 wells (rows A and B) with 100 μl of this mixture in each well. 2.4 ml evenly distributed between the wells is sufficiently accurate.
5. To the 1.6 ml of cell suspension remaining, add 2.4 ml of thymocyte medium.
6. Repeat steps 4 and 5 until all the wells are fitted and all the thymocyte medium used.
7. Top up each well with more growth medium.
8. Feed at day 5 and day 12 with two drops of medium.
9. Reassay supernatants.
10. Expand, freeze and propagate in mice, positive wells that appear monoclonal. If not monoclonal, reclone and repeat antibody screening test.

Comments:

Expect to have to clone approximately three times. The final clone plate, if originating from a single clone, will contain approximately 30% growth-positive wells, all of which are positive for antibody.

6.2.3 Cloning in soft agar

This procedure (Protocol 6.2.3) was that originally adopted by Kohler & Milstein (1975) and the details of the technique have been reviewed by Metcalf (1977). The resultant clones can be tested for antibody production following their transfer to liquid media, or alternatively it has been possible to demonstrate specific immunoglobulin secretion by clones directly in their solid agar growth support using techniques such as haemolytic overlays (Kohler & Milstein, 1975; Kohler, 1979), immunoprecipitation (Cook & Scharff, 1977) or replica immunoabsorption (Sharon et al, 1980). In general this method of cloning offers no advantages over the limiting dilution procedures and is considered by some to be less convenient.

PROTOCOL 6.2.3: CLONING IN SOFT AGAR

Materials:

Agarose solution (2% w/v in twice-distilled water in glass bottles, autoclave at 120°C for 15 min and store at 4°C)
RPMI growth medium, double strength with 20% FCS
24-well culture plates
Water bath at 44°C
Cells to be cloned

Method:

1. Melt the agarose in boiling water and allow to equilibrate to 44°C, together with the culture medium.
2. Mix equal volumes of agarose solution and culture medium and keep at 44°C.
3. Dispense 1 ml of this mixture into each well of the 24-well culture plate and allow to solidify. Prepare two agarose wells for each positive culture.
4. Prepare cell suspensions of 2×10^3/ml and 1×10^3/ml.
5. Mix 0.5 ml of each cell suspension with 1 ml of agarose/culture medium mixture.
6. Add 0.6 ml of each cell/agarose sample to a solid agarose well.
7. Allow this top layer to solidify and incubate the plate in a humid, 5% CO_2 incubator.
8. Within one to two weeks individual cell colonies will become visible as white dots. Assay for antibody secretion, either by using an overlay technique or by picking off individual colonies for liquid culture and subsequent assay of liquid supernatants (see Chapter 4).

Comment:

If cloning efficiencies are low, a feeder cell layer should be introduced under the agarose.

6.2.4 Cloning by fluorescence activated cell sorting (FACS)

A more sophisticated cloning method is offered by fluorescence activated cell sorting (FACS). Although a FACS machine will be beyond the reach of most readers, the method is worth mentioning since potentially hundreds of specific monoclonal cell lines can be cloned in one step (Herzenberg, 1978; Parks et al, 1979; Dangl & Herzenberg, 1982). Briefly, the procedure involves mixing and incubating together fluorescent antigen-coated latex microspheres (Polysciences, 0.9 μm diameter) and a hybrid cell mixture. The spheres bind exclusively to hybrids expressing the appropriate cell surface antibody, which therefore become fluorescently labelled, whilst irrelevant cells do not. The mixture may then be analysed cell by cell using a FACS apparatus (Beckton Dickinson FACS Division). Measurement of cell size, viability, and fluorescence can be made by the deflection of fine laser beams as a stream of single cell-containing droplets is passed through the machine. The desired cells can then be sorted by magnetic deflection of the appropriate droplets. Even single cells can be directed into individual wells of a 96-well culture plate containing feeder layers and then cultured in the usual way.

6.3 PROPAGATION OF SELECTED CLONES

6.3.1 Propagation in vitro

Once useful clones have been identified and isolated, and samples stored, the question then arises as to how these clones may be propagated in such quantity that useful amounts of monoclonal antibody may be derived from them. On the laboratory scale, cells can be grown in stationary culture to produce typical antibody levels in the supernatant of 5–50 μg/ml. Antibody yield is enhanced by growing the culture until the cell density is so great that cell death occurs. Similarly, such yields can be further improved if a proportion of the culture medium is frequently replaced with fresh medium. Scaling up can be facilitated by using spinner flasks or roller bottles of one-litre capacity. Antibody concentrations of 170 μg/ml or 34 mg/litre per day have been achieved in spinner flasks with semi-continuous feeding (Reuveny et al, 1986).

Fermenters, originally developed for microbial growth, have been adapted for the growth of hybridoma cells. Mammalian cells have more complex nutritional needs and tolerate much lower densities than microbial cultures. They therefore require a more carefully controlled culture environment. Hybridoma cells can either be grown in direct suspension or attached to microcarriers to increase the culture density. One of the major differences between types of fermenter or bioreactor is the method of keeping the cells or microcarriers in suspension. Air-lift fermenters operate by introducing gas mixtures into the base of the vessel which are directed upwards through a central draught cylinder and are dispelled at the surface (see Fig. 6.3.1a). The result is that medium and cells are circulated between the inner and outer compartments. This system avoids the potential shearing forces of stirred tank fermenters in which metal stirrers come into contact with the culture. Sizes of fermenters range from laboratory scale of five litres to commercial scales of 1000 and even 10 000 litres, and typical ranges of antibody yield can be between 40 and 500 mg/litre (Birch et al, 1985).

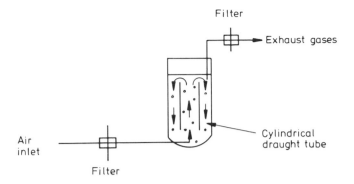

Cells are circulated gently through the medium by fine gas bubbles

Fig. 6.3.1a Schematic diagram of an air-lift fermenter

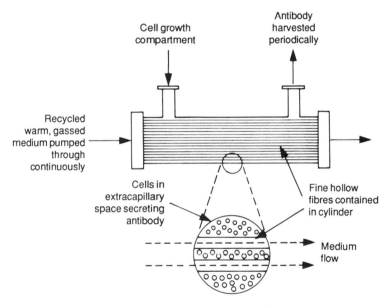

Fig. 6.3.1b Schematic diagram of hollow fibre reactor

Even higher yields of antibody can be achieved by adapting fermenters to continuous perfusion bioreactors. The fermenter is set up in circuit with a pump and filter, preferably of the hollow fibre type to prevent clogging. A proportion of the total medium volume is continuously replaced from a reservoir of fresh medium and a similar volume is harvested downstream from the filter. In comparative studies with fed-batch culture, antibody yields can be improved by at least five- to seven-fold (Microgon Technical Report; Reuveny et al, 1986). Nevertheless, these methods of propagation generate enormous volumes of culture medium and expensive sera, which will need concentration and further purification.

An increasingly popular method of producing high yields of antibody in vitro is to grow hybridoma cells within a hollow fibre reactor (see Fig. 6.3.1b). Several commercial companies are now marketing such systems with variable degrees of automation (and cost) (see Appendix B). The basic principle is that of perfusion of culture medium with necessary CO_2 buffering and temperature control, through a disposable hollow fibre cartridge in which the cells are growing. The cartridge is made up of hundreds of hollow fibres of sufficient pore size to allow medium and essential nutrients to pass through the fibres to the extracapillary space containing the growing cells. Secreted antibody, which cannot pass through the fibres, is harvested periodically from the extracapillary compartment over periods of months. It is claimed that the more sophisticated models can yield up to 20 g of antibody per month, depending on cell line characteristics. One drawback for research laboratories is the cost of the disposable hollow fibre cartridge which can be as much as £700. However, a recent report on the use of inexpensive hollow fibre reactors originally designed for haemodialysis claims antibody production of 30–200 mg per cartridge per day (Klerx et al, 1988).

It is essential with all cell propagation procedures that the fidelity of specific antibody productions be monitored regularly. This is necessary because of the ever-present risk that non-producing revertent cells may outgrow the specific antibody-producing cells. This instability varies from one cell clone to another. Some cultures may continue to yield specific antibody for months after cloning whereas others may be stable for much shorter periods. Such instabilities may not be too serious if fresh samples of the clones or recloned cells are available from storage and can be substituted for the cultures that no longer demonstrate the desired traits. If expected antibody levels in culture supernatant are diminishing, recloning the culture may be indicated.

6.3.2 Propagation in vivo

6.3.2.1 ASCITES PRODUCTION FROM RODENT HYBRIDOMAS

At the research laboratory level it is possible to obtain greater amounts (up to 1000-fold) of antibody quickly by adopting an in vivo approach to the propagation of hybrids (see Protocol 6.3.2.1). 10^6–10^7 cells may be injected intraperitoneally into pristane-treated histocompatible animals and the hybridoma allowed to grow for at least two weeks. Pristane is a branched-chain alkane and is the plasmacytogenic component of mineral oils which are known to induce plasmacytomas when injected intraperitoneally into mice (Potter & Robertson, 1960; Potter & Boyce, 1962; Anderson & Potter, 1969).

Pristane also acts by severely depressing the normal immunological function of the animal (Freund & Blair, 1982). The response to the cells will vary considerably according to the individual cell line, both in terms of volumes and concentrations produced and in the period of development. Normally, within two weeks of cell injection, ascitic fluid containing 5–15 mg/ml of antibody accumulates in the peritoneum from which it may be withdrawn with syringe and needle. Occasionally, solid tumours will develop instead of ascites. This can usually be overcome by incubating tumour cells in vitro for a few days and re-injecting them into more pristane-primed animals, though it is advisable to re-assay the culture supernatant before re-injection to ensure continued specific antibody production.

It is very important to use animals that are histocompatible with both parent cell lines (i.e. Balb/c mice or Lou rats). If another strain is used for the immune lymphocyte donor, it will be necessary to use F1 hybrid animals for ascites production.

PROTOCOL 6.3.2.1: PRODUCTION OF ANTIBODY IN ASCITIC FLUID

Materials:

  Pristane (0.5 ml per mouse; 1 ml per rat)
  Cyclophosphamide (1 mg/ml) for rats only
  PBS (0.5 ml per mouse; 1 ml per rat)
  Hybrid cells (minimum 10^6 per mouse; 10^7 per rat)
  Syringe, 21-g needle and small tube for aspirating

Method:

Pretreatment:

1. 10–60 days before cell injection, inject 0.5 ml of pristane i.p. into adult mice (1 ml for rats).
2. For rats it is necessary to inject the animals with 1 ml (1 mg) of cyclophosphamide the day before cell injection.

Cell injection:

3. Prepare healthily growing hybrid cells for injection by counting and centrifuging at 1500 rpm (400 g).
4. Resuspend pellet in appropriate volume of PBS (0.5 ml per mouse; 1 ml per rat) and inject i.p.

Fluid collection:

5. Within two weeks abdominal swelling will be observed. When the animal is the size of a normal pregnant female, fluid can be aspirated.
6. Hold the animal as for intraperitoneal injections and swab the abdomen with alcohol. Aspirate fluid from the peritoneal cavity with syringe and 21-g needle (or collect drops into small tube after puncture). 2–5 ml of fluid should be obtainable from a mouse (up to 15 ml from a rat). Kill the animal by cervical dislocation.
7. Spin ascitic fluid at 3000 rpm (1500 g) for 10 min. Store supernatant at $-20°C$ prior to testing.

Comments:

Whereas, in the past, fluid could be aspirated repeatedly over several days, the Home Office Inspectorate in the UK now allows for this to be done only once before sacrificing the animal.

Injecting larger numbers of cells will cause earlier induction of ascites but it will be more difficult to monitor progress. Sudden, unexpected deaths may occur before fluid can be harvested.

6.3.2.2 ASCITES PRODUCTION FROM HUMAN HYBRIDOMAS

Obviously, in vivo propagation of human hybridomas is not as straightforward as for rodent cells. However, ascitic fluid has been produced in Balb/c nude mice in the following way. Firstly, 10^7 hybrid cells are injected subcutaneously into irradiated (350 rad) Balb/c nude mice. The tumour is removed three to four weeks later and, after several passages of the cells in vitro, is re-injected into Pristane-primed irradiated Balb/c nude mice.

Chapter 7

Characterisation, purification and labelling

7.1 CHARACTERISATION

Once a cellular source of monoclonal antibody has been established in culture, it is usual to obtain a small quantity of ascitic fluid (i.e. from five animals) for further characterisation before preparing larger stocks of the antibody. A number of tests need to be carried out in order to relate the outcome to the final application and required specificity of the antibody. This chapter is concerned with the general characterisation of monoclonal antibody preparations, such as titre, isotyping, affinity and epitope analysis. This is followed by a consideration of antibody purification and labelling methods which will be of use in a variety of applications. In addition, these pilot studies will also include an assessment of the growth characteristics of the hybridoma, its stability on freezing, its suitability for the production of ascites and the suitability of the antibody for direct labelling. Clearly, however suitable the specificity of the antibody may be, it will be of limited use if the hybridoma grows poorly, has a low recovery from frozen stock or will not produce ascitic fluid.

7.1.1 Determination of titre

The titre of an antibody preparation is a measure of its concentration under a defined set of conditions. Thus, there will be a different measured titre for each assay system used. Indeed, some monoclonal antibodies, while of high titre in one assay system, may not react at all in another system.

In order to determine titre, serial dilutions of antibody preparation (e.g. pooled ascites, supernatant or IgG) are allowed to react with a known amount of antigen. The antibody titre is usually defined as the lowest dilution to bind significantly to antigen, or, alternatively, the dilution of antibody that gives a half-maximal binding to antigen. These definitions are illustrated in Fig. 7.1.1, which shows the results of an ELISA titre test as described in Protocol 7.1.1.

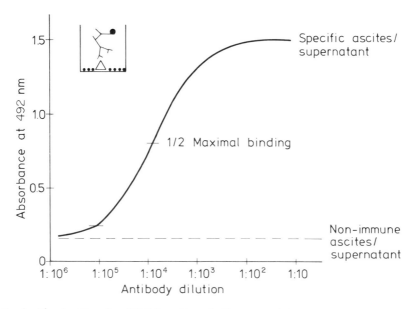

Fig. 7.1.1 Antibody titre by ELISA: either half-maximal binding, i.e. $1:10^4$, or minimal detectable response above negative control, i.e. $1:10^5$

PROTOCOL 7.1.1: ELISA: ANTIBODY TITRE TEST

This illustration is based on the adsorption of soluble antigen directly to the assay plate. It could equally well be adapted for use with indirectly adsorbed antigen or indeed other assay systems. More details of the practical aspects of each step in ELISA are given in section 4.4.1.1.1.

Materials:

Microtitre plate for ELISA
Antigen for coating plate
Irrelevant protein (i.e. 1% BSA) for negative control coating
Coating buffer
Wash buffer
Anti-mouse or rat Ig–enzyme conjugate
Enzyme substrate
Substrate stopping solution
Blocking solution
Monoclonal antibody (ascitic fluid, Ig preparation or culture supernatant) in serial 10-fold dilutions to zero
Negative control antibody as above (e.g. raised to an irrelevant cell line or to myeloma cells)
Multichannel pipette
Plate washer or wash bottle

Method:

1. Adsorb predetermined maximal concentration of antigen to four rows (48 wells) of the microtitre plate. Dilute in coating buffer and incubate 100 μl per well overnight at 4°C. Also coat two rows (24 wells) with irrelevant protein (i.e. 1% BSA in coating buffer) to serve as negative control.
2. Aspirate antigen solution and block unbound sites on the plastic by incubating 100 μl of blocking solution per well for 15 min.
3. Wash off unbound antigen and block protein three times with wash buffer.
4. Add 10-fold dilutions of ascitic fluid (or Ig or culture supernatant) across the plate to two of the antigen-coated rows and to the two negative control rows. Add the same dilutions of negative ascitic fluid to the other two antigen-coated rows as negative control. Incubate 100 μl per well for one hour at ambient temperature.
5. Wash off unbound antibody three times with wash buffer.
6. Add enzyme-labelled anti-mouse Ig (or rat or human as appropriate) at the recommended or predetermined dilution. Incubate 100 μl per well for one hour at ambient temperature.
7. Wash off unbound label three times with wash buffer.
8. Add enzyme substrate. Incubate 100 μl per well at ambient temperature until colour develops sufficiently (ideally approx. 30 min).
9. Do not wash off. Add 50 μl of the appropriate stopping solution and assess either visually or quantitatively in a spectrophotometer designed to read microtitre plates.

continued on next page

> *continued*
>
> *Comment:*
>
> If this assay was used as the basis for the antibody-screening test, the monoclonal antibody will, of course, be positive in this assay, obviating the need for a positive control antibody at stage 4. A positive control should be included if the titre is being tested in a different type of assay for the first time.

7.1.2 Determination of isotype

The class of antibody may have been selected for by the screening tests used during the establishment of the hybridomas. For example, since the majority of desired monoclonal antibodies will be IgG, if the anti-mouse whole Ig label in the test is replaced with an anti-mouse IgG label, only IgG antibodies will be selected. Similarly, if only a particular IgG isotype is required, the labelled anti-mouse IgG antibody can be replaced with anti-IgG$_1$, IgG$_{2a}$, IgG$_{2b}$, or IgG$_3$, all of which are available commercially. Equivalent anti-rat IgG and anti-human isotypes are also available. However, being so selective in the early screening tests may result in many potentially useful antibodies being missed. So, unless the search for a particular isotype is paramount, isotype testing should be confined to the later stages of antibody selection. In the later cloning stages, isotype analysis can also contribute to the decision on the likelihood of a particular clone being monoclonal.

Adaptation of the existing screening test to include isotype testing, as described above, may seem the simplest approach for the identification of an isotype in a specific case. However, isotype-specific reagents are expensive to buy individually and, although potentially thousands of antibodies can be tested, many research laboratories have no need for this capacity. It may, therefore, be more cost effective, in a research context, to use a commercially available isotype testing kit. Although such a kit will have a smaller capacity, it will contain all the necessary reagents, including standard controls for the isotyping of a number of monoclonal antibody products. Kits are available, based on several different assay systems, including ELISA, immunodiffusion and red cell agglutination.

Serotec market a kit based on red cell agglutination which is relatively cheap and only takes one hour to perform. The test is based on the fact that monoclonal anti-isotype-coated SRBCs agglutinate when in the presence of test antibody of the appropriate isotype. Kits for use with mouse and rat antibodies are available, and those for human antibodies are under development.

Alternatively, Amersham International market a kit which is a chromogenic assay based on a dipstick format. In this test a wide range of class- and subclass-specific antibodies, including $\varkappa$ and λ light chains and a positive control, are coated on discrete areas of the stick. The stick is then dipped sequentially into test antibody fluid, peroxidase-labelled anti-species antibody and substrate to produce a chromogenic reaction, thus enabling identification of isotype with the minimum of effort.

7.1.3 Cross-reactivity

An important early characterisation test of any panel of antibodies is the analysis of whether they react with the same, close or totally different epitopes. The simplest way of testing this characteristic is to label one of the antibodies directly and to allow it to compete for antigen with unlabelled antibodies of other sources as illustrated in Protocol 7.1.3. For this it is necessary to allow serial dilutions of unlabelled antibody (from saturating levels to zero) to bind to antigen, to wash off any excess and to apply the competing antibody label. If the label binding is maximal throughout the whole range of unlabelled antibody concentrations, it may be deduced that the two antibodies are directed towards different epitopes (see Fig. 7.1.3). The binding of labelled antibody may be either totally or partially inhibited. This will reflect the existence of competition between the antibodies for the same or very close epitopes or the fact that the binding of one antibody causes steric alterations in the molecule to prevent the other binding to its own distinct epitope. However, results should be interpreted carefully, since a high-affinity antibody label may displace a low-affinity antibody at the same epitope.

The possibilities of applying different antigen variations in the context of studying antibody–epitope cross-reactivity are endless. For instance, synthetic peptides altered by single amino acids may be employed to define the precise epitope to which a particular antibody binds. Similarly, other antigen variations could be studied, such as cells from different subsets, cultures, anatomical sites or species, chemically altered molecules, genetic variations in molecules that are responsible for certain disease states or virus species, etc.

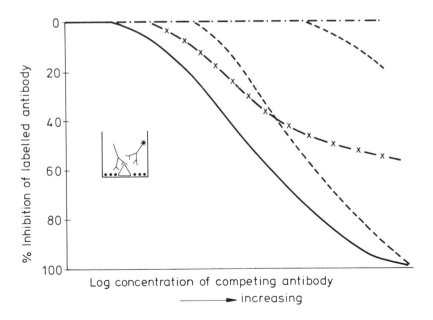

Fig. 7.1.3 Cross-reactivity test. Key: ———, inhibition by the same antibody; ---, competition by possibly lower affinity antibody at same or similar epitope; - · -, zero inhibition by unrelated antibody; -x-, maximal inhibition of label limited, so antibody probably directed at different though close epitope

PROTOCOL 7.1.3: ANTIBODY CROSS-REACTIVITY TEST

As an example, this protocol is based on an immunoradiometric assay (IRMA) which is more sensitive than ELISA, and the antibody under test can easily be labelled with ^{125}I (see section 7.4.1).

Materials:

   Standard buffers and equipment for IRMA
   Antigen-coated assay tubes
   ^{125}I-labelled test monoclonal antibody
   Unlabelled competing monoclonal antibodies
      (10-fold dilutions $1 : 100$–$1 : 10^7$, starting from a common protein concentration)

Method:

1. Coat assay tubes with predetermined maximal concentration of antigen, diluted in carbonate/bicarbonate buffer, pH 9.5.
2. Wash off excess antigen.
3. Incubate the tubes with serial dilutions (in duplicate) of unlabelled competing antibodies for one hour at 37°C. Include controls of unlabelled test antibody to demonstrate maximal competition and also totally unrelated antibody and buffer to demonstrate zero competition.
4. Wash off excess antibody.
5. Incubate tubes with labelled test antibody at 10 000 counts per tube for one hour at 37°C.
6. Wash off excess label.
7. Count tubes in gamma-counter for 1 min each.
8. Calculate the percentage inhibition of labelled antibody binding for each competing antibody dilution, based on the maximal counts obtained with 100% binding of labelled antibody. Plot these values against the log of the concentration of competing antibody to produce a series of curves as illustrated in Fig. 7.1.3.

7.1.4   Affinity

Affinity is a description of the strength of interaction between an antibody and an antigen in a given system. Interactions of high affinity (K values of 10^9–10^{12}) are essential for use in immunoassays where the bond between antigen and antibody has to withstand vigorous washing procedures. On the other hand, for immunopurification, low-affinity reactions (K values of 10^6–10^7) are preferred such that the bond between antigen and antibody can be easily broken when necessary without damaging the antigen. It is important, therefore, to choose from amongst the initial panel of monoclonal antibodies those with appropriate affinities, before finalising their application.

A very rough estimate of affinity can be deduced by comparing the Ig concentrations of antibody-containing culture supernatants with the strength of the responses measured in specific antigen assays. For instance, if there is a strong specific antibody response, but a relatively low general immunoglobulin level, the antibody–antigen reaction is likely to be of higher affinity than if the response is relatively weak.

In practice, obtaining an accurate measure of affinity is not easy. For soluble protein antigens that are easily purified, sufficient quantities may be available and consequently they can be labelled easily. In these situations, an RIA method with Scatchard analysis, described below (Protocol 7.1.4a), which enables affinity constants (K) to be calculated, may be appropriate. However, the slightest alteration in the antigen molecule caused by the labelling process may destroy the epitope to which the monoclonal antibody is directed, or at least alter the affinity of its binding. Thus data obtained using labelled antigen should be interpreted with caution. Conversely, an affinity constant obtained using free antibody may be changed when using a labelled antibody in an assay.

PROTOCOL 7.1.4a: DETERMINATION OF ANTIBODY AFFINITY (SCATCHARD ANALYSIS)

This protocol is based on the binding of serial dilutions of a radiolabelled antigen tracer to a known (predetermined) concentration of antibody and the separation of free from bound reactants by precipitation with anti-species antibody. A series of molar values is obtained for bound antigen from precipitated counts; the total counts and specific activity of the label are known, so the free labelled antigen can be calculated. The ratio of bound label to free (B/F) is plotted against bound label (B) to obtain a Scatchard plot and the affinity constant, K, is found from the slope.

Materials:

Antibody solution (predetermined concentration that binds 50% of label)
Labelled antigen (of known specific activity, ~50 000 cpm per tube)
Unlabelled antigen (10-fold serial dilutions, ~0.1–10^4 ng/ml for MW of 70 000, or molar equivalent for smaller peptides)
PBS, pH 7.4
Serum matrix (5% horse serum)
Sheep anti-mouse IgG (IgG fraction) bound to solid phase such as magnetic particles or cellulose or Sac-cel (Wellcome)
(1 ml per tube at second antibody excess, ~10 μg per tube)
Test-tubes
Gamma-counter

Method:

1. Predetermine the dilution of antibody that binds 50% of the labelled antigen, by measuring binding of a wide range of serial dilutions of antibody (i.e. 10×, 10-fold dilutions) to 50 000 cpm of label and precipitating bound complexes as described below. Ensure that a full sigmoid curve is obtained, since high-dose hook effects are common.

continued on next page

continued

2. Prepare serial dilutions (10×, 10-fold) of unlabelled antigen solution in PBS.
3. Dispense 100 μl of each dilution into assay tubes in triplicate (or even quintuplicate for greater accuracy).
4. Add 200 μl (50 000 cpm) of labelled antigen solution to each tube. Add 100 μl of serum matrix.
5. Add 100 μl of antibody solution.
6. Incubate overnight at room temperature.
7. Add 1 ml of second antibody/solid phase for precipitation to each tube and vortex-mix thoroughly. Alternatively, PEG could be used but may also precipitate free label.
8. Sediment precipitates by centrifugation.
9. Carefully decant supernates to waste.
10. Quantify radioactivity in pellets using gamma-counter.
11. Record measurements and analyse data as shown below.

Data analysis:

At equilibrium

$$[Ag] + [Ab] = [Ag : Ab]$$

where [Ag] is the molar concentration of free antigen
[Ab] is the molar concentration of free antibody
[Ag : Ab] is the concentration of antigen–antibody complex

$$\text{Affinity constant, } K = \frac{[Ag : Ab]}{[Ag] [Ab]} \text{ litres/mole}$$

1. The added label counts are known, and the specific activity of the label, so calculate the mass of added labelled antigen.
2. Calculate the total mass of antigen (cold and labelled) for each tube.
3. Calculate the mean counts of bound antigen (from the precipitated counts) and subtract the non-specific binding.
4. Calculate the percentage of the added label bound. Convert this value into mass.
5. Calculate the mass of free antigen.
6. Calculate the ratio of bound to free antigen.
7. Calculate the concentration of bound antigen in moles per litre, the reaction volume being 500 μl in this case.
8. Plot the ratio of bound antigen to free antigen $\frac{[Ag : Ab]}{[Ag]}$ against bound antigen [Ag : Ab].
A straight line of negative slope should be obtained (see Fig. 7.1.4a):

$$K = -\frac{1}{\text{slope}} \text{ or } -\frac{y}{x}$$

continued on next page

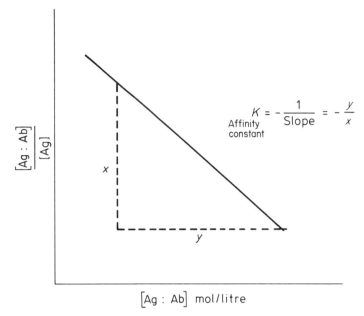

$$K = -\frac{1}{\text{Slope}} = -\frac{y}{x}$$

Affinity
constant

[Ag : Ab] mol/litre

Fig. 7.1.4a Scatchard plot

continued

Comments:

If a straight line is not obtained, the antibody may not be monoclonal (polyclonal sera gives curves, indicating a range of affinities). Alternatively, the labelling of antigen may have altered the epitope such that antibody is binding with different affinities in the labelled and cold preparations. This calculation assumes an antigen : antibody ratio of 1 : 1.

If the antigen is not available in sufficient quantities or in pure form, and cannot be labelled without alteration, the relative affinities of the antibody–antigen reactions for a panel of antibodies can still be estimated in a simple assay. For this, trace amounts of antigen are coated to a solid phase, then serial dilutions of antibody are allowed to bind and the bound antibody detected by using a labelled anti-species second antibody (Protocol 7.1.4b). A plot of the percentage of label bound versus the log of the antibody concentration will result in a series of curves (see Fig. 7.1.4b). At 50% maximal label binding the curve of the antibody with highest affinity will be on the left with decreasing affinities to the right. If the antibodies to be compared are of different isotypes, the use of Fab fragments would be better, to standardise the binding of the anti-Ig antibody.

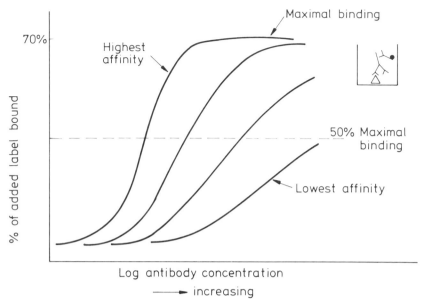

Fig. 7.1.4b Antibody affinity comparison. Maximal binding is never 100% because of impurities, free label or destroyed binding sites in the labelled preparation. Also some antibodies may never reach high enough concentrations in the supernatant to bind maximally

---

PROTOCOL 7.1.4b: AFFINITY RANKING

As an example, this test is based on ELISA but could equally well be adapted to other detection systems.

Materials:

Reagents and equipment for ELISA
Antigen for coating the solid phase
Serial 10-fold dilutions of antibodies to be compared (starting from a common protein concentration)
Anti-mouse IgG–HRP conjugate (or equivalent)

Method:

1. Coat the microtitre plate with a submaximal concentration of antigen in bicarbonate/carbonate buffer.
2. Wash off excess antigen.
3. Add serial dilutions of each competing antibody (in duplicate) and incubate for one hour at room temperature. Include buffer controls to obtain maximal label binding levels.
4. Wash off excess antibody.
5. Add anti-mouse IgG–HRP conjugate and incubate for one hour at room temperature.
6. Wash off excess conjugate.
7. Add enzyme substrate and stop the reaction when colour develops sufficiently.
8. Plot the percentage of added label that binds, against the concentration of antibody as illustrated in Fig. 7.1.4b.

continued on next page

continued

Comments:

Beware of variations in binding of the second antibody to different isotypes. If this is likely to be a problem, it may be worth preparing Fab fragments of the antibodies to be compared which can be detected using anti-mouse Fab enzyme conjugates.

7.2 ANTIBODY PURIFICATION

Ascitic fluid contains non-immune mouse IgG and most culture supernatant contains protein from bovine serum that might interfere with some applications of monoclonal antibodies. However, it is not always necessary to purify the antibodies, since they may be used at sufficient dilution and with appropriate controls to make the impurities irrelevant. Purification is necessary if the antibodies are to be labelled, since the greater the purity of the labelled antibody the greater the signal-to-noise ratio in the assay. Polyclonal antisera require affinity purification before labelling but this is not so critical with monoclonal antibodies. In practice, simple salt fractionation of ascitic fluid may be sufficient, although it may sometimes be necessary to protect the active site of the antibody with an immunoadsorbent. However, elution of the antibody from this adsorbent may cause additional damage, especially if the antibody is of very high affinity. The effect of the purification procedure on immunoreactivity should always be tested prior to use.

Two of the simplest antibody purification methods that do not require expensive hardware are described below in Protocols 7.2.1 and 7.2.2.

Greater levels of purity can be obtained, if desired, using a variety of other methods, including affinity chromatography (with protein A, antigen or antibody attached to the gel), ion exchange chromatography, FPLC and HPLC. Protein A-Sepharose can only be used with certain IgG subclasses that bind protein A. Of mouse IgG subclasses, IgG_{2b} binds to protein A with the highest affinity, followed by IgG_{2a}, and there is poor binding by IgG_1 and IgG_3 (Ey et al, 1978). Rat IgG_{2c} binds with the best affinity, followed by IgG_1, and there is low binding by IgG_{2a} and IgG_{2b} (Rousseaux et al, 1981). There are now a number of protein G products available, including a recombinant form (BDH) which binds strongly to IgG only of most mammalian species.

A simple, rapid method of IgG purification (Protocol 7.2.2) is chromatography on DEAE Affi-Gel Blue (Bio-Rad). This is a beaded crosslinked agarose with covalently linked cibacron blue F3GA and diethylaminoethyl groups. The blue dye displays differential affinity for several serum proteins. Application of a serum or ascitic fluid sample to an Affi-Gel column results in an immunoglobulin fraction purified of all major contaminants except transferrin. It is also free of any detectable proteolytic activity (including plasminogen). Yields of Ig are approximately 65%, at purities of over 95% with gradient elution, and 85–95% with two-step elution. A 7-ml column will process at least 1 ml of serum or ascitic fluid, in one procedure.

PROTOCOL 7.2.1: ANTIBODY PURIFICATION BY AMMONIUM SULPHATE PRECIPITATION

Materials:

Saturated ammonium sulphate (SAS):
    Dissolve 1000 g ammonium sulphate in one litre of H_2O at 50°C
    Stand overnight at room temperature.
    Adjust pH to 7.2 with dilute ammonia or H_2SO_4
Ascitic fluid
PBS
Ice bucket
Centrifuge tubes

Method:

1. Dilute ascitic fluid 1:2 in PBS on ice.
2. Add SAS dropwise to final concentration of 45% saturation (v/v)*, stir at 4°C for 30 min, and centrifuge at 1000 g, 15 min, 4°C.
3. Wash the precipitate with 45% SAS and recentrifuge at 1000 g, 15 min, 4°C.
4. Redissolve the precipitate in same volume of PBS as original ascitic fluid and centrifuge at 5000 g, 15 min, 4°C to remove any insoluble material.
5. Transfer supernatant to clean tube and reprecipitate Ig using final concentration of 40% SAS*.
6. Centrifuge at 1000 g, 15 min, 4°C.
7. Redissolve in minimum volume of PBS (0.5 ml) and dialyse against five litres PBS at 4°C overnight.
8. Centrifuge at 5000 g, 15 min, 4°C, to remove more insoluble material.
9. Test for protein content, aliquot and store supernatant at −70°C.

*To calculate the volume of salt solution (SAS) (ml) to be added per 1 ml of ascitic fluid sample, use the following equation:

$$\text{Volume (ml)} = \frac{1\ (SF - Si)}{1 - SF}$$

where SF = final saturation (as a fraction, not a percentage)
        Si = initial saturation

e.g. for 45%,   volume of SAS to add $= \dfrac{1\ (0.45 - 0)}{1 - 0.45} = 0.82\ \text{ml}$

to 1 ml ascitic fluid.

Calculation of protein content:

Prepare 1:20 dilution of purified antibody and determine absorbance at 280 nm in a spectrophotometer.

At 280 nm an OD of 1.0 (1-cm cuvette) = 0.69 mg/ml of mouse globulin, e.g. OD at 1:20 dilution = 0.95.
Protein concentration = 0.95 × 0.69 × 20 mg/ml.

**PROTOCOL 7.2.2: ASCITIC FLUID PURIFICATION USING
DEAE AFFI-GEL BLUE** (Bio-Rad; Bruck et al, 1982)

Materials:

7 ml DEAE Affi-Gel column
>1 litre column buffer/dialysis buffer:
  (0.02 M Tris-HCl, pH 7.2; see Appendix A)
30 ml Ig elution buffer:
  as above with 50 mM NaCl (2.9 g/l)
Gradient maker (optional)
Fraction collection tubes
Dialysis tubing
PEG, MW 40 000 for dry dialysis
One litre PBS
Reagents for column reconstitution (see below)

Method:

1. Prepare a 7-ml Affi-Gel column, according to the manufacturers' instructions, in a 10-ml syringe (prepacked columns can also be obtained) and equilibrate the column with column buffer.
2. Prepare ascitic fluid sample by dialysis against column buffer, followed by centrifuging at 10 000 g for 15 min to remove any fibrin precipitate. Keep sample at 4°C to minimise protease activity.
3. Apply 1 ml of sample to the column, and elute by using either a linear gradient of approximately 200 ml Tris-HCl buffer with 0–100 mM NaCl or a two-step gradient of 20 ml of column buffer to remove transferrin, followed by 30 ml Tris-HCl with 50 mM NaCl to collect the Ig fraction (15×2-ml fractions).
4. Measure the absorbance of each fraction at 280 nm in a spectrophotometer. The first eight fractions of the protein peak should contain immunoglobulin.
5. Calculate the protein concentration of the fractions, i.e. at 280 nm, an OD of 1.0 = 0.69 mg/ml mouse protein (Ig).
6. Pool the fractions and concentrate to 1 ml by dry dialysis against PEG, MW 40 000.
7. Finally dialyse the sample against PBS, add 1 M NaCl to bring the salt concentration to 150 mM and add 0.02% NaN_3 for storage at −70°C.
8. Reconstitute the column with approximately 50 ml of 2 M guanidine HCl or 8 M urea in 1.4 M HCl, followed by 100 ml of 0.5 M NaCl buffer and finally 50 ml of column buffer. The column can be reused at least 10 times without loss of efficiency.

Purification by HPLC is one of the most efficient and rapid methods of purification, although the initial equipment and column costs can be high. It also offers the most efficient method of idiotype separation, which is a useful analytical facility when constructing hybrid monoclonal antibodies. Various columns are available and are continuing to be developed. Detailed descriptions of their use in purifying monoclonal antibodies can be found in Burchiel et al (1984) and Juarez-Salinas et al (1984), and also commercial literature is a good source of information on the latest developments.

It should be noted that all the methods mentioned above, except antigen affinity chromatography, may copurify non-specific IgG or protein that may be present in FCS or in ascitic fluid. If antigen is not available in sufficient quantities or purity for affinity chromatography, it may be worth considering growth of the hybridoma in serum-free medium, to reduce this problem.

In general, IgM antibodies are less easily purified than IgG. An initial precipitation by prolonged dialysis against distilled water, pH 5.0, followed by preparative polyacrylamide electrophoresis, is a suggested approach.

7.3 ANTIBODY FRAGMENTATION

When assaying antigens that may contain Fc receptors, such as membrane antigens of lymphoid or reticuloendothelial origin, the use of whole antibody labels may result in unacceptably high levels of 'non-specific binding'. It may then be necessary to use antibody preparations in which the Fc portion has been removed. Fragmentation of immunoglobulin molecules can be achieved by controlled proteolysis, although the conditions will vary considerably with species and immunoglobulin class or subclass. Bivalent F(ab')$_2$ fragments are usually preferable, although more difficult to prepare, to the univalent Fab or Fab' fragments which may exacerbate problems with low-affinity antibodies.

7.3.1 Fab

Fab fragments can be obtained by proteolysis of the whole immunoglobulin by the non-specific thiol protease, papain (Stanworth & Turner, 1978; Protocol 7.3.1). Papain must first be activated by reducing the sulphydryl group in the active site with cysteine, 2-mercaptoethanol or dithiothreitol. Heavy metals in the stock enzyme should be removed by chelation with EDTA (2 mM) to ensure maximum enzyme activity. Enzyme and substrate are normally mixed together in the ratio of 1 : 100 by weight. After incubation, the reaction can be stopped by irreversible alkylation of the thiol group with iodoacetamide or simply by dialysis. The completeness of the digestion should be monitored by SDS-PAGE and the various fractions separated by protein A-Sepharose or ion exchange chromatography.

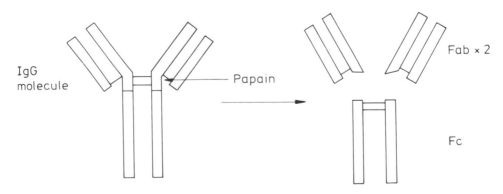

IgG
molecule

Papain

Fab × 2

Fc

Fig. 7.3.1 Fab

PROTOCOL 7.3.1: PREPARATION OF Fab FRAGMENTS OF IgG

Materials:

Purified IgG (20–30 mg in 4 ml)
5 mM phosphate buffer, pH 8.0
EDTA (Na$_2$) 26 mg/ml (100 μl)
Cysteine 63.2 mg/ml (100 μl)
Papain 10 mg/ml (30 μl)
DE52 cellulose
Na$_2$HPO$_4$.2H$_2$O (44.5 g/500 ml)
5–300 mM gradient phosphate buffer, pH 8.0 for chromatography

Method:

1. Incubate IgG with 100 μl of EDTA (final concentration 2 mM), 100 μl cysteine (final concentration 10 mM) and 30 μl papain 1 mg/100 mg IgG) for four hours at 37°C.
2. Stop the reaction by dialysis against 5 mM phosphate buffer.
3. Separate Fab fragments from Fc by ion exchange chromatography.
4. Prepare ion exchange column by equilibrating 40 g DE52 cellulose with 5 mM phosphate buffer and adjust pH back to 8.0 by adding Na$_2$HPO$_4$.2H$_2$O stock buffer. Wash off supernatant after cellulose settles and replace with half the gel volume of 5 mM phosphate buffer and degass. Pack 2-cm diameter column at the rate of 40 ml/hour and elute with two to three litres of buffer.
5. Apply Fab digest to column and elute with a gradient of 5–300 mM phosphate buffer, pH 8.0.
6. Two distinct peaks should be obtained, the first of which contains Fab. Pool fractions, concentrate, test and store.

Comments:

For mouse IgG$_{2a}$ and IgG$_{2b}$, 10 mM cysteine or 1 mM dithiothreitol is more than sufficient but IgG$_1$ is more resistant. All rat IgGs can be cleaved easily by 10 mM cysteine (two to four hours at 37°C) (Rousseaux et al, 1980, 1983, 1986).

If immunoglobulin digestion is incomplete, whole IgG will be found in the second peak. Test by antigen binding detected by anti-IgG label. An increase in reducing agent concentration should improve digestion, although too much may cause more extensive digestion of inter-chain disulphide bonds and subsequent loss of antibody activity.

7.3.2 F(ab')$_2$

The usual procedure for preparation of F(ab')$_2$ fragments from IgG of rabbit and human origin is limited proteolysis by the enzyme pepsin (Protocol 7.3.2). The conditions first described by Stanworth & Turner (1978) (100×antibody excess w/w in acetate buffer at pH 4.5, 37°C) suggest that antibody is cleaved at the C-terminal side of the inter-heavy-chain disulphide bond. Application of this method to mouse and rat IgG, has, however, met with some problems. Lamoyi & Nisonoff (1983) and Lamoyi (1986) found that rates of digestion of mouse IgG varied with subclass and that it was difficult to obtain high yields of active F(ab')$_2$ fragments without some undigested or completely degraded IgG. In particular, IgG$_{2b}$ was highly susceptible to complete degradation. The other subclasses required different incubation conditions to produce optimal results (see protocol). A study by Parham (1983) also expressed difficulty in obtaining F(ab')$_2$ fragments from IgG$_{2b}$ with pepsin. In addition, there were problems with cleavage of IgG$_1$ and IgG$_{2a}$, but these were overcome to some extent by changing the digestion conditions.

Digestion of rat IgG by pepsin also requires modification; Rousseaux et al (1983) found that rat IgG$_{2c}$ was most susceptible to cleavage. The conditions were dialysis in 0.1 M acetate buffer, pH 4.5, and then incubation for four hours with 1% w/w pepsin; IgG$_1$ and IgG$_{2a}$ digestion was improved if first dialysed against 0.1 M formate buffer, pH 2.8, at 4°C, for 16 hours followed by acetate buffer. IgG$_{2b}$ gave the most inconsistent results but incubation in staphylococcal V8 protease (3% w/w) in 0.1 M sodium phosphate buffer, pH 7.8, for four hours at 37°C, proved successful.

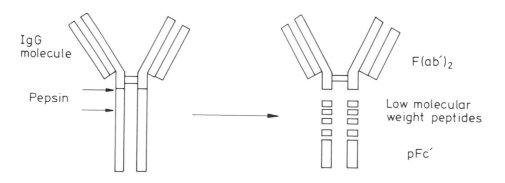

Fig. 7.3.2 F(ab')$_2$

PROTOCOL 7.3.2: PREPARATION OF F(ab')₂ FRAGMENTS OF MOUSE IgG
(Lamoyi & Nisonoff, 1983; Lamoyi, 1986)

It is important to know the subclass of antibody before attempting this preparation, since the experimental conditions will vary accordingly. Mouse IgG_{2b} is highly susceptible to complete breakdown by this method and the yields of fragments from the other subclasses are variable. Refer to the text for special conditions for rat IgG fragmentation.

Materials:

Purified IgG (1–10 mg/ml)
0.1 M sodium acetate buffer, pH 7.0 (for dialysis)
Pepsin (2 mg/ml in acetate buffer at appropriate pH for subclass)
PBS (for dialysis and gel filtration)
2 M Tris-HCl, pH 8.0
0.5 M NaOH
Sephadex G-150 column (90×2.5 cm)

Method:

1. Dialyse the IgG against 0.1 M sodium acetate buffer overnight at 4°C.
2. Adjust the pH according to the antibody subclass with 2 M acetic acid, dropwise and with stirring; pH 4.2 for IgG_1 and IgG_{2a}, 4.5 for IgG_3. Warm to 37°C.
3. Add pepsin to give an antibody excess of 33× and incubate for the following times according to subclass; eight hours for IgG_1, four to eight hours for IgG_{2a}, and 15 min for IgG_3. The expected yields of F(ab')₂ for the three subclasses are about 70%, 25–50% and 60% respectively.
4. Stop the reaction by adding 2 M Tris-HCl, pH 8.0 (1/40 volume) and raise the pH to 8.0 with 0.5 M NaOH.
5. Centrifuge and discard any precipitate.
6. Separate the fragments by gel filtration on Sephadex G-150 equilibrated with PBS.
7. Test digestion on SDS-PAGE.

Comments:

Individual IgG molecules may show different susceptibilities to digestion so, if this protocol yields disappointing results, the optimal conditions will need to be found experimentally.

7.4 LABELLING OF MONOCLONAL ANTIBODIES

Monoclonal antibodies can undoubtedly be valuable reagents in many immunoassays without extra purification or direct labelling. Indeed, the combination of monoclonal antibodies and indirect labels such as anti-species immunoglobulin may have more universal application, especially in the early stages of characterising panels of antibodies. Direct labelling, however, can improve the sensitivity of assays and takes full advantage of the unique specificity of monoclonal antibodies. The remainder of this chapter describes the relative merits of different labels and methods of directly labelling monoclonal antibodies.

7.4.1 Radiolabelling

The handling of radiochemicals necessitates extreme care and, because of the volatile nature of the elements should be performed in an efficient and safe fume hood designated for the purpose. Disposal of radioactive waste must be performed strictly according to the instructions of the site radiation protection officer and exposure levels of personnel should be monitored by film badge and thyroid scan.

The choice of isotope will, of course, depend on the proposed assay and the counting facilities available. The most common radioisotope used for labelling soluble proteins is ^{125}I, which has a half-life of 60 days. Labelled antibodies of high specific activity can be obtained by a variety of different methods which are commonly used to label antibodies of polyclonal origin. Monoclonal antibodies are particularly sensitive to the potentially damaging effects of iodination processes, so the milder methods may give more consistent results. It would be unusual, however, if none of the following methods were successful.

7.4.1.1 ^{125}I: CHLORAMINE T METHOD

Antibodies can be radioiodinated with ^{125}I to high specific activity by the chloramine T method (Greenwood et al, 1963; Protocol 7.4.1.1). The main amino acid to be labelled is tyrosine, although histidine may be used if the pH is greater than 8.0–8.5. Chloramine T is an oxidising agent, high concentrations of which can damage antibodies, so only 1–10 μg per reaction is necessary. Recommended antibody concentrations are 1 mg/ml or more in Tris-HCl or PBS at neutral pH. The reaction is usually stopped by adding a reducing agent such as sodium metabisulphite but this could also cause damage. An alternative is addition of 1 mM tyrosine, which has no potential damaging effects. Addition of NaI at the end helps to terminate the reaction and to minimise adsorption of radioiodide to the glassware. Fifty to ninety per cent of added ^{125}I should be incorporated into antibody (one iodine per IgG = approx. 12 μCi/μg).

PROTOCOL 7.4.1.1: ^{125}I ANTIBODY LABELLING: CHLORAMINE T METHOD
(after Greenwood et al, 1963)

Prepare all reactants in appropriate pipettes before starting the reaction, in an LP3 polystyrene tube with stopper, in a designated fume hood.

Materials:

LP3 reaction tube
Vortex mixer
100 μg IgG
14 MBq Na^{125}I
10 μl chloramine T (5 mg/ml)
100 μl sodium metabisulphite (1.2 mg/ml)
1 ml potassium iodide (200 μg/ml)
Sephadex G-25, 25-cm column
0.1 M PBS, 0.1% BSA, pH 7.4, as elution buffer
Rack of fraction tubes

Method:

1. Mix continuously on vortex mixer (in stoppered tube), IgG, Na^{125}I and chloramine T for 10 s.
2. Stop the reaction by adding sodium metabisulphite for a further 30 s.
3. Add 1 ml of potassium iodide to absorb free ^{125}I.
4. Separate iodinated antibody from unbound ^{125}I by gel chromatography. Collect 1-ml fractions.
5. Screen column fractions for radioactivity.
6. Pool void volume fractions of the first peak and store at 4°C until tested in test assay.
7. Dispose of other fractions and Sephadex according to local radiochemical handling regulations.

Comments:

A number of contaminants are capable of inhibiting chloramine T iodination, including reducing agents such as azide and thiocyanate ions. A modification of this method using chloramine T covalently bound to polystyrene beads (Iodo-beads) is apparently much gentler (Markwell, 1982).

7.4.1.2 ^{125}I: LACTOPEROXIDASE METHOD

A milder method of iodination involving incorporation of ^{125}I into tyrosine is the lactoperoxidase method (Marchalonis, 1969; Protocol 7.4.1.2). This depends on the use of H$_2$O$_2$ as an oxidising agent, the concentration of which is very critical. Too high a concentration may damage the enzyme and too low a concentration will reduce the

---

PROTOCOL 7.4.1.2: ^{125}I ANTIBODY LABELLING:
LACTOPEROXIDASE METHOD (adapted from Thorell & Johansson, 1971)

Materials:

LP3 polystyrene tube
18–66 MBq Na^{125}I (8–15 μl)
250 μg IgG (15–30 μl)
4 μg lactoperoxidase (2 μl of 1 mg/0.5 ml Sigma L2005 lyophilised), 1 μg 0.88 mM
    H$_2$O$_2$ (1 : 10 000 of 30% w/v solution)
PBS
Vortex mixer
Sephadex G-25, 25-cm column
0.1 M PBS, 0.1% BSA, pH 7.4, as elution buffer
Rack of fraction tubes

Method:

1. Mix continuously for 1 min the Na^{125}I, IgG, lactoperoxidase and H$_2$O$_2$ in the
   stoppered LP3 tube.
2. Stop the reaction by dilution with 0.5 ml of PBS.
3. The reaction mixture is then treated as in the chloramine T method, i.e.
   transferred to a Sephadex G-25 column equilibrated with PBS and saturated
   with 1% BSA in PBS.

Comments:

Any azide in the buffers or IgG preparation would inhibit the iodination.

---

specific activity of the labelled product (typically 30–70% of added counts). Also, the
antibody itself may be damaged by the H$_2$O$_2$. If concentration adjustment of H$_2$O$_2$
proves unsuccessful, one can use glucose oxidase to convert glucose into small steady
amounts of H$_2$O$_2$ in the reaction mixture (Hubbard & Cohn, 1975). 'Enzymobeads'
(Bio-Rad), coated with glucose oxidase and lactoperoxidase, are made for this purpose.

7.4.1.3 ^{125}I: IODOGEN METHOD

The iodogen method (Fraker & Speck, 1978; Protocol 7.4.1.3), also based on tyrosine
iodination, is potentially as efficient a method as the chloramine T method. It is,
however, less likely to damage antibodies, since they are not exposed to soluble
oxidising agents.

PROTOCOL 7.4.1.3: ^{125}I ANTIBODY LABELLING: IODOGEN METHOD
(after Fraker & Speck, 1978)

Materials:

Iodogen (1,3,4,6-tetrachloro-3a, 6a-diphenylglycoluril)
   20 µg/ml in methylene chloride (20 µl/0.4 µg per tube)
Plastic reaction tube (10 × 75 mm)
Dry nitrogen
Ice
2 mg/ml of Ig solution (50 µl)
11 µg/ml of KI (10 µl)
14 µCi Na^{125}I
Borate saline buffer (pH 8.3, ionic strength 0.1 M) (100 µl)
1 mM tyrosine
Sephadex G-25, 25-cm column
0.1 M PBS, 0.1% BSA, pH 7.4, as elution buffer
Rack of fraction tubes

Method:

1. Prepare iodogen-coated tubes by adding 20 µl iodogen to test tube (10 × 75 mm plastic) and evaporate the methylene chloride by gently blowing off in a stream of dry nitrogen to leave a thin film in the bottom of the tube. Such tubes can be stored in the dark at room temperature for several weeks.
2. Place an iodogen-coated tube on ice and add 50 µl of Ig solution.
3. Initiate the reaction by addition of 10 µl of KI and 14 µCi Na^{125}I.
4. Adjust the volume to 100 µl by adding borate saline buffer.
5. Incubate for 5 min with stirring and terminate the reaction by decanting the protein solution and adding an excess of 1 mM tyrosine.
6. Separate the labelled protein from free iodine by gel filtration, pool fractions and store as for the chloramine T method.

7.4.1.4 ^{125}I: BOLTON–HUNTER METHOD

If tyrosine is a component of the antigen-binding site of the antibody, then all the iodination methods described above will seriously affect immunoreactivity if the site is left unprotected by an immunoadsorbent. The Bolton–Hunter reagent (N-succinimidyl 3-(4-hydroxy-5-[^{125}I]iodophenyl) proprionate) labels the ε-amino group of lysine residues (Bolton & Hunter, 1973) by a gentle method, since no oxidising agents are involved (Protocol 7.4.1.4). The charge of the antibody is reduced by one unit for each modified lysine, so if the labelled antibody is to be applied in experiments based on charge, such as electrophoresis, there may be problems. Other disadvantages of this method are expense and the high susceptibility of the reagent to hydrolysis.

PROTOCOL 7.4.1.4: ^{125}I ANTIBODY LABELLING: BOLTON–HUNTER METHOD (after Bolton & Hunter, 1973)

Materials:

Dry iodinated Bolton–Hunter reagent
5 µg IgG in 10 µl 0.1 M borate buffer, pH 8.5
Ice bucket
0.5 ml 0.2 M glycine in 0.1 M borate buffer, pH 8.5
Sephadex G-25
0.05 M phosphate buffer, pH 7.5, containing 0.25% (w/v) gelatin

Method:

1. To the Bolton–Hunter reagent (purchased as a dry iodinated ester in a reaction vial), add 5 µg of IgG in cold borate buffer, and agitate in ice bucket for 15 min.
2. Stop the reaction by adding glycine solution for 5 min on ice.
3. Purify the product by gel filtration (Sephadex G-25, equilibrated with 0.05 M phosphate buffer with gelatin), pool fractions and store as for the chloramine T method.

Comments:

The use of Tris buffers containing free amines or azide ions inhibit the reaction. Also avoid the presence of albumin and other serum proteins, which bind the low molecular weight labelled products, until after the separation step.

7.4.1.5 ^{3}H

Other isotopes have in the past been less favourable as antibody labels because of the low specific activities obtainable. A relatively new procedure for labelling with ^{3}H based on reductive methylation has been introduced (Tack et al, 1980) which can label antibodies to high specific reactivities, of high stability and with minimal effects on protein configuration and charge. Tritium is also suitable for biosynthetic labelling of antibody which is described in section 7.4.6 and Protocol 7.4.1.

7.4.2 Fluorescent labelling

7.4.2.1 FITC AND TRITC

The most popular fluorochromes used in immunohistochemistry are fluorescein isothiocyanate (FITC) and tetramethyl rhodamine isothiocyanate (TRITC). They give contrasting green and red fluorescence and have frequently been made use of in two-colour staining techniques. FITC is also ideal for use in flow cytometry, where the 488-nm wavelength of the laser is close to the maximum excitation wavelength of FITC (495 nm). In this application, antigen-coated fluorescent microspheres containing hundreds of FITC molecules have proved to be more sensitive (Parks et al, 1979). TRITC is less sensitive in flow cytometry since its maximum excitation is at 554 nm. The difference in excitation wavelengths between these two fluorescent probes has restricted their simultaneous use in flow cytometry. However, this problem has been overcome by the use of dual laser systems.

Antibody-labelling procedures with FITC or TRITC usually involve simple dialysis of the antibody preparation against the fluorochrome solution, based on a competing hydrolysis (Protocol 7.4.2.1). However, obtaining the right ratio of fluorochrome to antibody and removal of unreacted components is critical for good resolution. Under-labelling of antibody will result in low-intensity fluorescence and, unless removed, unlabelled antibody will compete for antigenic determinants. Too much labelling will give rise to highly charged acidic species which will bind non-specifically to many preparations. Labelling of relatively impure antibody preparations can also result in non-specific binding due to other immunoglobulins or proteins with possible receptor sites on the tissue (i.e. Fc or transferrin receptors). Aggregates or complexes of immunoglobulins should be removed from conjugates immediately before use, by centrifugation (40 000 g, 30 min). Problems with Fc receptors can be overcome by labelling Fab or F(ab')$_2$ antibody preparations.

PROTOCOL 7.4.2.1: FLUORESCENT LABELLING OF ANTIBODY
(after Goding, 1976b)

Antibody concentrations of 10 mg/ml and more, relative to fluorochrome concentrations of 10–20 μg/mg of antibody, favour conjugation to antibody (up to 70%), but lower antibody concentrations (1 mg/ml) can be used if the fluorochrome concentration is correspondingly increased (100 μg/mg antibody).

Materials:

    Carbonate/bicarbonate buffer at pH 9.5, one litre (see Appendix A)
    IgG (1–20 mg/ml)
    Fluorochrome (1–10 mg/ml)
    Sephadex G-25
    0.02 M PBS, pH 7, with 0.1% NaN$_3$

Method:

1. Dialyse IgG sample overnight against the carbonate/bicarbonate buffer. This helps to remove extraneous nucleophiles such as Tris, amino acids, ammonium ions or azide that might inhibit conjugation. At this pH, a large fraction of the ϵ-amino groups on lysine are unprotonated, enabling thiourea bonds to form with the thiocyanate groups of the fluorochrome.
2. Prepare a stock solution of fluorochrome 1–10 mg/ml dissolved in DMSO to avoid hydrolysis.
3. Add the appropriate volume dropwise to the antibody solution and leave for two hours in the dark at room temperature.
4. Separate free fluorochrome from the conjugate by chromatography on Sephadex G-25 in 0.02 M PBS, pH 7, with 0.1% NaN$_3$. The first coloured band eluted is conjugated antibody.
5. Store at 4°C in the dark.
6. Test before use for F : P ratio (see below) and specificity of staining.

continued on next page

continued

Comment:

An alternative procedure based on FITC bound to a dispersing agent, Celite (available as a complex from Calbiochem), was reported by Rinderknecht (1962).

Calculation of F : P ratio:

The F : P ratio for fluorescein conjugates can be calculated using a simple formula proposed by The & Feltkamp (1970a,b). They include a correction for the contribution of FITC to the total absorbance at 280 nm.

$$\text{F : P ratio} = \frac{2.87 \times OD_{495}}{OD_{280} - 0.35 \times OD_{495}}$$

$$\text{Mouse [IgG]} = \frac{OD_{280} - 0.35 \times OD_{495}}{0.69} \quad \text{mg/ml}$$

A similar formula cannot be applied to TRITC conjugates, since they are more hydrophobic and therefore more likely to precipitate. Ratios can be roughly estimated from the ratio of the absorbance at 555 nm to the absorbance at 280 nm. High-ratio TRITC conjugates are susceptible to self-quenching due to rhodamine–rhodamine interactions (Goding, 1976b). In general, a molar ratio of FITC to antibody of 3 : 1 is suitable for tissue sections and 5–6 : 1 for cell suspensions. TRITC conjugates with a ratio of 1 : 2 are usually optimal.

7.4.2.2 PHYCOBILIPROTEIN

Recently, phycobiliproteins have been introduced as fluorescent dyes (Oi et al, 1982; Kronick & Grossman, 1983).

They have a broad excitation range and a very large extinction coefficient compared to small organic fluorochromes. A subgroup, the phycoerythrins, have major excitation peaks between 490 nm and 560 nm, with emission at 570 nm. They can, therefore, be excited by the same laser as fluorescein, giving orange fluorescence in contrast to the green of fluorescein.

Production of phycobiliprotein–antibody conjugates is based on covalent linkage similar to those applied in enzyme–antibody conjugation. The two most successful linkages appear to be either a disulphide bond using SPDP (*N*-succinimidyl 3-(2-pyridylthio)-proprionate) or a thioether bond using SMPB (succinimidyl 4-(*p*-maleimidophenyl)-butyrate).

7.4.3 Enzyme labelling

In choosing an enzyme with which to conjugate an antibody, one of the first considerations is the proposed assay system. Those criteria, for instance, that are most suitable for ELISA are not necessarily the best for immunohistochemical methods. Endogenous enzyme or other substances in the sample to be assayed could interfere with the specific result. So too, could assay conditions (pH, ionic strength, buffer composition etc.) which are incompatible with the sample or antibody–antigen reaction. The properties of an ideal enzyme label include a high turnover between substrate and product, easily detectable activity, purity and low cost of enzyme, low K_m for substrate, high K_m for product, and high K_i. Safety should also be a consideration, since many substrates are carcinogenic or mutagenic. The substrate and its product should be soluble in ELISA-type assays to keep light scattering to a minimum but, in immunohistochemistry or immunoblotting, the product must be insoluble such that it is deposited and remains as near as possible to the site of the antigen.

There are many different methods of conjugating antibody to enzyme. The criteria to consider are high yield up to 100%, with a well-defined product composition, minimal, if any, inactivation of either enzyme or antibody, stability on storage, and ease and cost of preparation. As with the choice of enzyme, different assays have different requirements in terms of enzyme to antibody ratios. In immunohistochemical assays a 1 : 1 ratio is preferred, since this will penetrate tissues easier. ELISAs, however, require the highest enzyme to antibody ratio to amplify the signal, as long as activity is not impaired. The success of a particular method will depend on the relative concentrations of reactants, relative reaction rates of reactants with crosslinking reagent, ionic strength and pH of buffers.

It is beyond the scope of this book to explore all the choices available, but suggested approaches in two assay systems (ELISA and immunohistochemistry) are given below. An excellent source for more detailed information is Tijssen (1985).

7.4.3.1 HORSERADISH PEROXIDASE

The most universal enzyme to be applied in immunoassays is horseradish peroxidase (HRP), for which there is a wide variety of suitable substrates, yielding soluble or insoluble products. HRP activity is measured indirectly by the rate of transformation of a H-donor. The concentration of H_2O_2 is critical due to its small optimum range and, at higher concentrations, its inhibitory effect on the substrate. It is also unstable in storage.

7.4.3.1.1 HRP substrates

The favoured H-donors for producing soluble products suitable for ELISA are 5-AS, ABTS, OPD and TMB. Detailed recipes for these are given in Appendix A.

5-AS (5-aminosalicylic acid) after purification (pure form not available commercially; see Ellens & Gielkens (1980)) yields a very soluble, very stable product with low background with an absorbance at 500 nm.

ABTS (2,2'-azino-di-(3-ethyl-benzthiazoline sulphonate-6)) (Childs & Bardsley, 1975; Porstmann et al, 1981) is colourless, but on oxidation gives a blue, rather unstable product with an optimum absorbance at 414 nm; however, it is potentially mutagenic.

OPD (o-phenylenediamine) has an orange oxidation product which can be measured at low concentrations at 492 nm. This is probably the most sensitive activity indicator but is light sensitive and possibly mutagenic.

TMB (3,3,5,5'-tetramethylbenzidine) appears not to be as hazardous and can be as sensitive as OPD. The reaction gives a soluble blue or yellow product absorbing at 655 or 450 nm (Bos et al, 1981).

Stable, insoluble products of HRP substrates are required in immunohistochemistry and immunoblotting.

DAB (3,3'-diaminobenzidine) (Graham & Karnovsky, 1966) is most often used, although there is indirect evidence that it may be carcinogenic. The colour reaction can be further enhanced by the application of cobalt, nickel (Adams, 1981) or metallic silver (Gallyas et al, 1982).

Alternative substrates are Hanker–Yates reagent (p-phenylenediamine-HCl and pyrocatechol), AEC (3-amino-9-ethylcarbazole) (Graham et al, 1965), 4-CN (4-chloro-1-naphthol) (Nakane, 1968), and α-naphthol/pyronin (Tubbs & Shebani, 1981; Sofroniew & Schrell, 1982).

7.4.3.1.2 HRP–antibody conjugation

The recommended methods of conjugating HRP to antibody are the periodate method (Nakane & Kawaoi 1974; Protocol 7.4.3.1.2) or CHM-NHS (4-(N-maleimidoethyl)-cyclohexane-1-carboxylic acid N-hydroxysuccinimide ester) (Yoshitake et al, 1982; Ishikawa et al, 1983; Protocol 7.4.3.2.1). The periodate method couples through the carbohydrate portion of each molecule and, since this is not usually involved in the active site of the antibody or enzyme, the method is less likely to reduce immunoreactivity. CHM-NHS is a heterobifunctional agent based on maleimide crosslinking of enzyme to thiolated antibody. It yields largely monoconjugates of enzyme to antibody with an efficiency in the region of 75%. The two-step glutaraldehyde method (Avrameas & Ternynck, 1971), linking the ε-amino groups of lysine residues, is often favoured for its simplicity and mild conditions but is said to be inferior to the periodate method in terms of yield (Voller et al, 1979) and detectability of conjugates (Nygren, 1982).

PROTOCOL 7.4.3.1.2: HRP CONJUGATION: MODIFIED PERIODATE METHOD (after Wilson & Nakane, 1978)

Materials:

HRP (4 mg)
Fresh 100 mM $NaIO_4$ (0.2 ml)
1 mM sodium acetate buffer, pH 4.4 (for dialysis)
200 mM carbonate buffer, pH 9.5 (20 μl)
8 mg IgG (dissolved in 1 ml of 10 mM carbonate buffer, pH 9.5)
Fresh $NaBH_4$ solution (4 mg/ml in water) (0.1 ml)
Con A-Sepharose affinity column
Con A buffer (see Appendix A)
10–100 mM α-methyl-D-mannopyranoside in Con A buffer

Method:

1. Dissolve 4 mg HRP in 1 ml distilled H_2O.
2. Add 0.2 ml fresh 100 mM $NaIO_4$ and stir for 20 min at room temperature. The carbohydrate groups on the enzyme are oxidised by the $NaIO_4$.
3. Dialyse overnight at 4°C against 1 mM sodium acetate buffer, pH 4.4. This pH reduces the likelihood of auto-conjugation.
4. Add 20 μl 200 mM carbonate buffer, pH 9.5, 8 mg IgG (dissolved in 1 ml of 10 mM carbonate buffer, pH 9.5) and stir for two hours at room temperature. The aldehyde groups on the activated enzyme are now free to react with the amino groups of IgG.
5. The mixture is stabilised by reduction in 0.1 ml fresh $NaBH_4$ solution (4 mg/ml in water) for two hours at 4°C.
6. Remove free peroxidase from the conjugate mixture by adding an equal volume of saturated ammonium sulphate, incubating for one hour, centrifuging at 6000 g for 15 min, and rewashing with half-saturated ammonium sulphate. The free peroxidase remains in solution.
7. Remove free IgG by affinity chromatography on Con A-Sepharose which binds peroxidase and peroxidase conjugates. Dialyse conjugate in Con A buffer, apply to column and allow passage of free IgG (monitored at 278 nm). Peroxidase conjugate can then be desorbed with 10–100 mM α-methyl-D-mannopyranoside in Con A buffer.
8. Store the conjugate at −70°C with stabilising proteins (i.e. 1% BSA) or add an equal volume of glycerol and store at −20°C.

7.4.3.2 OTHER ENZYMES

Alternative enzymes used in ELISA include β-galactosidase (β-G), alkaline phosphatase (AP) and urease. A suitable soluble substrate for β-G is *o*-nitrophenyl-β-D-galactosidase (ONPG) (Appendix A), which yields a yellow product absorbing at 410 nm. For AP, the usual substrate is *p*-nitrophenyl phosphate (p-NPP) in a diethanolamine buffer

Table 7.4.3.2   Antibody–enzyme conjugation methods

Enzyme	Conjugation method	References
HRP	NaIO$_4$ CHM-NHS Two-step glutaraldehyde	Wilson & Nakane (1978) Yoshitake et al (1982) Avrameas & Ternynck (1971)
AP	One-step glutaraldehyde OPDM CHM-NHS	Avrameas (1969) Kato et al (1975) Hamaguchi et al (1979)
β-G	OPDM MBS	Kitagawa & Aikawa (1976) O'Sullivan et al (1979)
Urease	One-step glutaraldehyde	
Glucose oxidase	CHM-NHS	

(Appendix A). This also yields a yellow product absorbing at 405 nm. The coloured products of β-G and AP are less pronounced than those of HRP but can be useful if background staining of HRP poses problems, particularly in plant tissue. Both enzymes are more expensive but β-G is more efficiently conjugated than HRP and is capable of detecting smaller amounts of antigen. As for AP, the conjugation methods and detectability are inferior to the HRP equivalent. A common mistake, however, with AP conjugates is the use of PBS as a wash buffer, since as little as 1% of inorganic phosphate significantly inhibits AP.

If endogenous enzyme presents problems in detection, urease has an advantage in being absent from mammalian cells, and produces a clear end-point of yellow to purple when in the presence of the pH indicator bromocresol purple (urease releases ammonia from the substrate urea) but there can be a rapid loss of enzyme activity.

Alternative enzymes to HRP that can be used in immunohistochemistry and immunoblotting include alkaline phosphatase (Mason & Sammons, 1978) and glucose oxidase (Rathlev & Franks, 1982). β-Galactosidase is unsuitable because of its size. Some of the more popular conjugation methods can be successfully applied to several different enzymes. The recommended methods for each enzyme mentioned above are summarised in Table 7.4.3.2, together with source references. Two of the methods, CHM-NHS and two-step glutaraldehyde, are described in Protocols 7.4.3.2.1 and 7.4.3.2.2 respectively.

PROTOCOL 7.4.3.2.1: ENZYME LABELLING OF ANTIBODY: CHM-NHS METHOD

Recommended for glucose oxidase, alkaline phosphatase, and peroxidase (but not β-galactosidase, for which the OPDM method is better).

Materials:

    2 mg enzyme
    100 mM sodium phosphate buffer, pH 7.0 (0.3 ml)
    1.6 mg CHM-NHS in 20 μl dimethylformamide (DMF)
    Sephadex G-25
    100 mM phosphate buffer, pH 6.0 (column buffer)
    Fab' 1–4 mg/ml in column buffer with added 5 mM EDTA or thiolated IgG
    1 ml 2-mercaptoethanol (2-ME)

Method:

 1. Dissolve enzyme in 0.3 ml of 100 mM phosphate buffer, pH 7.0.
 2. Add CHM-NHS in DMF (precipitation can be prevented by first warming DMF to 30°C), stir and incubate at 30°C for one hour.
 3. Centrifuge to remove precipitate.
 4. Apply supernatant to Sephadex G-25 column equilibrated and eluted with phosphate buffer, pH 6.0.
 5. Pool enzyme-containing fractions (measure protein absorbance at 280 nm) and concentrate by dry dialysis or equivalent.
 6. Add Fab' or thiolated IgG so that the final concentration of enzyme and antibody is 2–6 mg/ml.
 7. Incubate overnight at 4°C (or one hour at 30°C).
 8. Add 1 ml 2-ME to block remaining maleimide groups.
 9. Purify conjugates by gel filtration (i.e. Ultrogel AcA 44 column (LKB)) or affinity chromatography.
10. Store by adding an equal volume of glycerol and keep at −20°C.

Comments:

Fab' gives a higher yield than whole IgG, which needs to be thiolated (see method of O'Sullivan et al (1979)). NaN$_3$ should not be used to preserve conjugates since it decomposes maleimide.

PROTOCOL 7.4.3.2.2: ENZYME LABELLING OF ANTIBODY: TWO-STEP GLUTARALDEHYDE METHOD (after Avrameas & Ternynck, 1971)

Materials:

Enzyme (10 mg)
100 mM phosphate buffer, pH 6.8
Glutaraldehyde (GA)
Sephadex G-25
0.9% NaCl for gel column
IgG (5 mg) or Fab' (2.5 mg)
500 mM sodium carbonate buffer, pH 9.5
1 M lysine, pH 7.0 (0.1 ml)
PBS
0.22 μm filter

Method:

1. Incubate 10 mg of enzyme in 0.2 ml 100 mM phosphate buffer, pH 6.8, and excess GA for 18 hours at room temperature.
2. Remove excess GA by gel filtration on Sephadex G-25 equilibrated with 0.9% NaCl (or dialyse).
3. Concentrate enzyme–GA to 1 ml.
4. Add antibody (5 mg IgG or 2.5 mg Fab'/ml of 0.9% NaCl) and 0.2 ml of 500 mM sodium carbonate buffer.
5. Incubate overnight at 4°C and block remaining activated groups with 0.1 ml 1 M lysine for two hours.
6. Dialyse overnight against PBS, filter-sterilise, purify and store as described in the preceding enzyme conjugation protocols.

7.4.4 Chemiluminescent labelling

Non-isotopic immunoassays based on the use of antibodies labelled with chemiluminescent molecules are beginning to find favour as highly sensitive, stable and safe alternatives to radioimmunoassays (Woodhead et al, 1981; Weeks & Woodhead, 1984). Chemiluminescence occurs when the vibronically excited product of an exoergic chemical reaction reverts to the ground state with the emission of photons. Photon emission can then be detected by relatively simple equipment (luminometers; see Appendix B).

The chemiluminescent molecules that have been used for protein labelling are luminol (5-amino-2,3-dihydrophthalazine-1,4-dione) or its structural variants or acridinium esters.

7.4.4.1 LUMINOL

Light emission from luminol is derived from the addition of alkaline hydrogen peroxide and a catalyst such as horseradish peroxidase or cytochromes or even simple transition metal cations. The reaction is complex, requiring careful controls, and luminol itself can lose considerable chemiluminescent activity when coupled to proteins. However, one advantage of this system is that peroxidase bound to antibody in an ELISA can be used to catalyse the chemiluminescent reaction, thus leading to signal enhancement of the standard ELISA (Thorpe et al, 1985).

Of the luminol derivatives, 7-N-(4-aminobutyl-N-ethyl)-naphthalene-1,2-dicarboxylic acid hydrazide (ABEN) is the most appropriate for antibody labelling in terms of light output and favourable coupling conditions (Schroeder et al, 1978). The coupling first involves preparation of the hemisuccinamide of ABEN using succinic anhydride and then formation of an active ester with N-hydroxysuccinamide and dicyclohexylcarbodiimide. The active ester is then incubated with antibody (ratio 20 : 1, ester : antibody), resulting in two to five molecules of label incorporated per antibody molecule.

PROTOCOL 7.4.4.2: ACRIDINIUM ESTER LABELLING OF ANTIBODY
(after Weeks et al, 1983)

Materials:

Acridinium ester (AE; see text)
(100 μg in 400 μl of dry acetonitrile)
50 μl of purified IgG
200 μl of 0.1 M phosphate buffer, pH 8.0
100 μl of lysine monohydrochloride (10 mg/ml in pH 8.0 phosphate buffer)
Sephadex G-25 (medium grade)
Elution buffer: PBS, pH 6.3, containing 0.01% sodium azide, 0.1% human serum
albumin, and 20 mg/l normal horse IgG
0.05 M NaOH containing 0.05% H_2O_2 (100 volumes)

Method:

1. Place 10 μl of the dissolved AE preparation in a glass vial.
2. Dissolve 50 μl of purified IgG in 200 μl of 0.1 M phosphate buffer, pH 8.0, and add to the acridinium solution.
3. Leave to stand for 15 min at room temperature.
4. Add 100 μl of lysine monohydrochloride (10 mg/ml in pH 8.0 phosphate buffer) and leave for a further 15 min.
5. Apply to a column of Sephadex G-25 (medium grade), eluted with PBS, pH 6.3, containing 0.01% sodium azide, 0.1% human serum albumin, and 20 mg/l normal horse IgG.
6. Collect 500-μl fractions and test 10-μl (diluted 1 : 100 in elution buffer) samples for luminometry.
7. In the luminometer, add 200 μl of 0.05 M NaOH containing 0.05% H_2O_2 (100 volumes) to the 10 μl of diluted column fraction. The emitted counts are integrated over 10 s.
8. Pool the void volume fractions containing the peak of chemiluminescent activity and store in aliquots at $-20°C$.

Comment:
Excess label incorporation can result in loss of immunoreactivity of the antibody. A ratio of up to three label molecules to each antibody is recommended.

7.4.4.2   ACRIDINIUM ESTER

Acridinium esters have been shown to be more sensitive labels than the luminol derivatives and have distinct advantages based on the mechanism of chemiluminescent emission. The excited product, *N*-methyl acridone, is cleaved from the parent compound by mild oxidation and no catalysts are required. The emission is then independent of any possible chemical modifications in coupling label to protein and is less susceptible to background and interference effects.

The first acridinium ester to be used for immunolabelling was 4-(2-succinimidyl-oxycarbonylethyl)-phenyl-10-methylacridinium-9-carboxylate fluorosulphonate (AE), prepared from either acridine or diphenylamine via acridine-9-carboxylic acid (Weeks et al, 1983). It reacts with primary and secondary aliphatic amines through an amide linkage (thus the use of amine-containing buffers such as Tris-HCl should be avoided). The coupling reaction can be carried out rapidly under mild conditions and the conjugate separated by gel filtration as described in Protocol 7.4.4.2.

7.4.5   Gold labelling

The use of colloidal gold probes was first introduced by Faulk & Taylor (1971) in the field of immunoelectronmicroscopy and is now widely used also in light microscopic immunohistochemistry and immunochemistry (Roth, 1983; Moeremans et al, 1984). They are also used as tracers in binding, uptake or transport studies where probes as small as 5 nm can be followed in living cells. Gold particles are very electron dense and so are easily distinguished from biological structures in electron microscopy. They can generate both secondary and back-scattered electron images, which is ideal in scanning electron microscopy. In immunoblotting, direct antibody–gold probes produce a red colour without the need for further development as with other labels. If necessary, low concentrations can be further enhanced by precipitation of larger silver particles on the gold surface (Moeremans et al, 1984). Similarly, in light microscopy, visualisation of small, well-penetrating colloidal gold particles can be enhanced (Holgate et al, 1983).

Gold particles of various sizes (5, 10, 15, 20, 30 and 40 nm) can be obtained commercially (Appendix B) ready for either direct labelling to antibody or to protein A or secondary antibodies. These size variations can be exploited in double-staining techniques to localise two or more different antibodies. Colloidal gold binds macromolecules by non-covalent electrostatic adsorption which stabilises the gold against electrolyte-induced flocculation. Preparation of gold probes is relatively simple, provided optimal conditions are first determined. These will depend on the gold particle size, ionic concentration, concentration and molecular weight of the protein, and pH, which should be close to the p*I* of the protein. Details of gold probe preparations including specific monoclonal antibody applications are provided by the gold particle manufacturers.

7.4.6   Biotin labelling

Many assay methods are able to take advantage of the signal-enhancing biotin–avidin system (Wilchek & Bayer, 1984; Guesdon et al, 1979; Kendall et al, 1983). Biotin

---

PROTOCOL 7.4.6: BIOTIN LABELLING OF ANTIBODY
(after Guesdon et al, 1979)

Materials:

IgG (10 mg/ml)
0.1 M NaHCO₃, pH 8.0–8.3 (for dialysis and column buffer)
0.1 M (34.1 mg/ml) solution of BNHS or 45.4 mg/ml of BX-NHS dissolved in DMSO (or dimethylformamide)
Sephadex G-25 column if required

Method:

1. Dialyse the purified antibody preparation overnight against 0.1 M NaHCO₃, pH 8.0–8.3, and adjust to a concentration of 10 mg/ml in the same buffer.
2. For each 1 ml of antibody solution, add 57 μl of a 0.1 M (34.1 mg/ml) solution of BNHS (45.4 mg/ml of BX-NHS) dissolved in DMSO (or dimethylformamide) immediately before use (the BNHS should be allowed to warm before opening the bottle to reduce the risk of hydrolysis).
3. Leave at room temperature for one hour.
4. Dialyse overnight against several changes of 0.1 M NaHCO₃ at 4°C, or by chromatography of a Sephadex G-25 column equilibrated in 0.1 M NaHCO₃, to separate the conjugate from the low molecular weight reaction by-products.
5. Store with 0.1% thimersal at 4°C or an equal volume of glycerol at −20°C.

---

conjugated to antibody can be detected by avidin or streptavidin conjugated to a variety of different labels or stains, many of which are available commercially. In this way, a greater number of label molecules can be attached indirectly to the antibody than is possible by direct labelling. Non-specific reactions can be reduced and, if used in conjunction with a contrasting direct label, two monoclonal antibodies from the same species can be simultaneously detected.

The usual procedure for conjugating antibodies with biotin is similar to that described for fluorochrome labelling (Protocol 7.4.6). Biotin can be obtained commercially in the form of biotinyl-N-hydroxysuccinimide ester (BNHS, Sigma) or biotinyl-ε-aminocaproic acid N-hydroxysuccinimide ester (BX-NHS, Calbiochem) or synthesised by the method of Jasiewicz et al (1976). The nucleophilic unprotonated ε-amino groups of lysine on the protein bind to the activated ester group to form an amide bond, with the release of N-hydroxysuccinimide. Less biotin substitution per antibody molecule can be achieved with BX-NHS compared to BNHS without loss of assay sensitivity. As with fluorochrome labelling, the reaction is in competition with a hydrolytic reaction which is dependent on the antibody concentration. At protein concentrations of 10 mg/ml, biotin binding is favoured, with hydrolysis taking over as the protein concentration is reduced.

7.4.7 Biosynthetic labelling

Production of monoclonal antibodies in vitro offers unique opportunities to label them internally, by incorporating radiolabelled amino acids into antibody synthesis. Although the products are usually of limited specific activity, no chemical manipulation of antibody, including prior purification, is necessary. The procedure is very simple (Protocol 7.4.7), involving incubation of the appropriate hybridoma cells in a medium deficient in a particular amino acid, with the addition of that amino acid in radiolabelled form. The labels most often used are ^{3}H- or ^{14}C-labelled lysine, arginine, phenylalanine or leucine, and [^{35}S] methionine. Lysine is favoured since the number of amino acid residues is usually higher. ^{3}H-labelling is preferred for quantitative assays and immunocytochemistry because of its higher specific activity. Labelling with ^{125}I is not advisable for safety reasons. Neither is the use of multiple labelled amino acids, which can adversely affect the protein-synthesising capacity of the cells. The choice of amino acid has been discussed in more detail by Vitetta et al (1976), Galfre et al (1981) and Coligan et al (1983).

PROTOCOL 7.4.7: BIOSYNTHETIC LABELLING WITH ^{3}H

Materials:

2×10^6 log phase hybridoma cells
Labelling medium: lysine-free tissue culture medium (i.e. Gibco SelectAmine kit) containing 5% dialysed serum and $40 \, \mu g/ml$ [^{3}H]lysine
24-well culture plate

Method:

1. Wash hybridoma cells in medium deficient in the labelled amino acid.
2. Remove the supernatant completely and resuspend cells in 1 ml of labelling medium in a well of a 24-well plate.
3. Incubate overnight at 37°C in a humid 5% CO_2 incubator.
4. Centrifuge (150 g for 10 min) and remove the supernatant.
5. Store at -20°C.

Comment:

About 70% of the label should be monoclonal antibody. Further purification may not be necessary for SDS-PAGE analyses but extensive dialysis against distilled water should be sufficient to reduce background binding in more quantitative assays.

Chapter 8

Human hybridomas and engineered antibodies

8.1 HUMAN HYBRIDOMAS

Monoclonal antibodies of rodent origin have been widely applied in many scientific and medical fields, as research tools and in diagnostic tests in particular. However, their application in therapeutics and in vivo diagnosis has been hampered by the risks of allergic reactions and the neutralisation of any potentially beneficial effects produced in response to the foreign protein. These adverse effects could be avoided by using antibodies of human origin, which have the added advantage of being more effective at harnessing human effector systems than rodent antibodies.

Considerable efforts have been made, therefore, to produce human monoclonal antibodies. In addition to the therapeutic advantages, human antibodies can probe human immune responses, e.g. in autoimmune diseases and cancer, and there may be specificities in man not readily available in xenogeneic immunisations. However, producing human antibodies has not been as straightforward as was the case with the original mouse system and it still remains a difficult objective.

The major practical problems include immunisation, the source of immune lymphocytes and the limitations of the immortal fusion partner. The process of active immunisation can, of course, be a major problem in humans, and this is reflected by the fact that most of the human antibodies that have been made have been derived either from patients naturally infected with bacteria or viruses, from approved immunisation schedules or from patients with active disease. As a result, more attention has been paid to in vitro immunisation methods for the production of human monoclonal antibodies. Many different strategies have been reported which are far from being standardised; however, some are now beginning to yield results. The main problem seems to be in the use of lymphocytes from peripheral blood, which, of course, is the readily accessible source in humans. In general, spleen or tonsil lymphocytes make better fusion partners than do cells from peripheral blood. The probable reason for this is that, in peripheral blood, there are not enough antigen-reactive specific B-cells in an appropriate state of differentiation and proliferation. Furthermore, there are more suppressor and cytotoxic cells in the peripheral circulation and not enough antigen-presenting cells. A number of different strategies have been employed to enrich the antigen-specific B-cell population and to deplete cytotoxic T-cells in such sources. These methods have been described in detail in Chapter 5.

In the mouse system, good myeloma lines were developed soon after the hybridoma method was first reported and there was little need for individual laboratories to develop their own. As a result, most mouse hybridomas available today are derived from only a few myeloma lines. The opposite is true in the human system, where there are many different fusion partners, none of which are truly satisfactory. True human myelomas grow poorly in culture and EBV-transformed lymphoblastoid cell lines tend not to be in the most appropriate stage of development for high fusion efficiency.

Efforts to produce human monoclonal antibodies from alternative fusion partnerships have included the fusion of human B-lymphocytes directly with mouse myelomas or with heteromyelomas created from the fusion of two or more human and mouse myelomas. As described in Chapter 5, human monoclonal antibodies can also be produced by activation and transformation of lymphocytes by Epstein–Barr virus (EBV), without the involvement of any fusion techniques. Better results are obtainable, however, by applying EBV transformation in combination with conventional fusions, either with mouse or human partners. A summary of the development of human cell lines is given in Chapter 5, together with recommendations on their use.

Despite the low efficiency of the available immunisation and immortalisation procedures, human monoclonal antibodies have been produced, although thorough evaluation of the different methods has yet to be done. Even so, the generally low levels of antibody synthesis and frequent loss of this function continue to be a problem. In most cases the reasons for this have not been fully investigated. It may be proposed, however, that loss of expression may be a result of either the loss of structural genes, defects in transcription or translation or a failure in assembly and secretion. If the reason for the loss of antibody secretion were known accurately, then it might be possible to take corrective measures. More information on the technical aspects of human monoclonal antibody production can be found in Olsson et al (1983), Carson & Freimark (1986), Brown (1987) and Abrams et al (1986). Comprehensive reviews of the human monoclonal antibodies that have been produced have been published by O'Hare & Yiu (1987) and James & Bell (1987).

8.2 HYBRIDOMAS OF OTHER SPECIES

A small number of heterohybridomas have also been formed for the production of monoclonal antibodies of species other than rodent or human, such reagents being particularly appropriate for use in the veterinary and food production fields. Immune lymphocytes from rabbit, cow, pig, sheep, rhesus monkey and chimpanzee have all been used (Yarmush et al, 1980; Raybould et al, 1985a,b; Guidry et al, 1986; Groves et al, 1987; Van Meurs & Jonker, 1986; Van Meel et al, 1985). The other fusion partner has usually been one of the mouse myeloma lines but, as with human–mouse heterohybridoma production, low fusion efficiencies and poor stability remain the major problems. More success has been achieved using second-generation heterohybridomas, where the original hybridoma between mouse and the other species is rendered HAT sensitive so that it can be used as a fusion partner in a refusion with lymphocytes from the other species (Tucker et al, 1984; Anderson et al, 1986).

8.3 BISPECIFIC ANTIBODIES

Bispecific monoclonal antibodies are immunoglobulin molecules in which each of the two antigen-binding regions are of a different specificity. They are derived from fusions between a specific antibody-secreting hybridoma and a lymphocyte secreting a different antibody. Milstein & Cuello (1983) were the first to take this approach by fusing a rat anti-peroxidase hybridoma with spleen cells from a rat immunised with somatostatin. The methodology is the same as described in previous chapters, with the requirement to create a HAT-sensitive derivative of the parent hybridoma, so that post-fusion hybrids can be selected in HAT medium. The resultant hybrid hybridomas secrete antibodies with dual specificity which in this case were applied in immunohistochemistry. (See Fig. 8.3.)

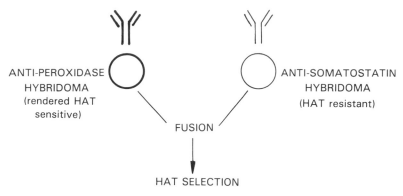

ANTI-PEROXIDASE
HYBRIDOMA
(rendered HAT
sensitive)

ANTI-SOMATOSTATIN
HYBRIDOMA
(HAT resistant)

FUSION

HAT SELECTION

8 different hybrid hybridomas produced secreting different
combinations of light and heavy chains + 2 species of unfused hybridomas

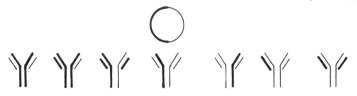

1 : 8 will be producing antibody of the desired combination

Fig. 8.3 Hybrid hybridomas (as described by Milstein & Cuello, 1983)

Although the process of production of these antibodies takes some considerable time, the potential for a successful outcome is greater than with chemical modification methods which risk denaturation and resulting loss of antibody activity. In addition, the use of anti-peroxidase bispecific antibodies in immunohistochemistry eliminates the risk of non-specific binding by secondary antibodies or enzyme complexes. The molecules are also smaller, so penetration is improved in techniques such as electronmicroscopy. However, such antibodies do exhibit one disadvantage in that the binding affinities of the bispecific antibody are likely to be weaker than those of the individual parent antibodies. Also, a hybrid hybridoma cell has the capacity to produce 10^8 different antibody molecules resulting from recombination of all the different heavy and light chains, since the antibody genes of both parents are codominantly expressed. Thus, the bispecific antibody will need to be purified from the other possible antibody constructs including bivalent monospecific antibodies from unfused hybridomas. It should also be borne in mind that any monospecific antibodies are likely to inhibit binding of the bispecific antibodies in the screening tests.

A number of authors have used this technique to harness cytotoxic T-lymphocytes (CTL) and target them to MHC-negative cells that normally evade the T-cell response (known as effector cell targeting). For example, Lanzavecchia & Scheidegger (1987b) created bispecific antibodies from fusions between a HAT-sensitive anti-CD3 hybridoma and spleen cells from mice immunised with either EBV-transformed human B-cells, human red blood cells or ovary carcinoma cells.

Two hybridoma lines can also be fused, if selection of the resulting hybrid hybridoma is made possible by rendering one of the parent lines susceptible to inhibition of a different metabolic pathway. Lanzavecchia & Scheidegger (1987b) produced a HAT-sensitive, G-418-resistant (see Chapter 5) anti-CD3 hybridoma. Thus, when this is fused with the usual HAT-resistant hybridoma, both parental cell lines will die in medium containing HAT and G-418, whereas the hybrids will survive. Alternative metabolic inhibitors that have been used for this purpose are emetine (an inhibitor of protein synthesis blocking the translocation step) and actinomycin D (an inhibitor of RNA synthesis) (Suresh et al, 1986).

In applying these antibodies to therapeutic practice, there are a number of complications. F(ab')$_2$ fragments should be used, since whole antibody molecules can bind to Fc receptors on some cells, and therefore form a bridge between these cells and the CTLs. Any bivalent anti-CD3 antibodies remaining in the reaction mixture should also be removed to avoid the risk of reciprocal killing of CD3-positive effector cells (Lanzavecchia & Scheidegger, 1987b). If administered in vivo, the full effects of the free bispecific antibody might not be exploited due to the small number of CTLs in the appropriate state. However, this problem might be overcome by stimulation of CTLs in vitro, by anti-CD3 and IL-2, followed by incubation with hybrid antibody before administration in vivo.

A more universal application of bispecific antibodies in effector cell targeting was proposed by Gilliland et al (1988) which obviates the need for individual antibodies to be made for each target cell specificity. These authors fused two hybridomas, one producing antibodies specific for CD3 and the other for rat Igk-1b allotype. In this way, any target cell, regardless of cell surface antigen, which can be coated in vitro with the rat allotype, can be caused to lyse when in the presence of bispecific antibody and effector cells.

8.4 ENGINEERED ANTIBODIES

The problems encountered in applying conventional hybridoma technology to the human system have given encouragement to the use of molecular biological methods in the production of human monoclonal antibodies. Clearly, the production of monoclonal antibodies of particular specificities, affinities and isotype, is a relatively imprecise process using hybridoma methodology when compared with the potential offered by the rapidly developing techniques of genetic engineering.

Early attempts at altering antibody function were made through somatic mutation. This process occurs spontaneously in myeloma or hybridoma cells (sometimes resulting in loss of antibody expression) but it can be promoted by mutagenesis. Class switch variants arising from such conditions have been isolated, enabling alteration of effector functions in antibodies of particular specificities (Radbruch et al, 1980; Neuberger & Rajewsky, 1981). Such mutations, however, depend on the chance occurrence of the desired changes in the protein.

By contrast, gene transfection into lymphoid cells provides a means of producing immunoglobulin molecules that can be altered in a defined way. In order to understand how these changes can be effected by gene manipulation, there follows a brief description of the origins of antibody diversity. More detailed information can be found in standard immunology textbooks (i.e. Roitt et al, 1985; Weir, 1986) and reviews such as Honjo (1983), Wall & Kuehl (1983) and Milstein (1987).

8.4.1   Origins of antibody diversity

The complete variable regions of immunoglobulin genes are made up of several fragments, V,D (diversity region), J (joining region), for heavy chains, and V,J for light chains. In germ line DNA, there are several hundred V_H and V_L genes, twelve D genes, and four J_H, four J_x and two J_λ genes (the latter two giving rise to x and/or λ light chain variation). The many different combinations of these genes provide the basic level of diversity (see Fig. 8.4.1). Additional diversity is generated by the imprecision of these recombinational events and also somatic point mutations, to which the variable region may be particularly susceptible. The diversity is then further amplified by the different potential combinations of heavy and light chains.

Constant region genes (C) are responsible for immunoglobulin isotype variation and are located downstream from the J segment genes, in the order μ, δ, $\gamma3$, $\gamma1$, $\gamma2b$, $\gamma2a$, ϵ, α in the mouse. Each of these C genes contains exons coding for each domain of the heavy chain. At the 5' end of each C gene (except δ) there is a corresponding switching sequence. This enables the recombination of any C gene directly with the VDJ region. During the primary antibody response, the VDJ region is transcribed with a μ gene, resulting in the production of IgM antibodies. During B-cell maturation, however, another C gene (i.e. γ) takes the place of the μ gene at its switch region and the intervening genes are probably lost. The result is typical of the secondary response in that the secretion of IgG antibodies is predominant. The exons coding for these genes are separated by intron sequences which facilitate their rearrangement both in vivo and by conventional recombinant DNA technology in vitro.

8.4.2   General methods

8.4.2.1   GENE CLONING

Sufficient is known about mouse and human immunoglobulin gene sequences (Kabat et al, 1987) to enable the reconstruction and cloning of gene fragments. Immunoglobulin gene libraries for mouse and human are generally available on application to the originating laboratories. Messenger RNA can be purified from hybridoma cells (Saiki et al, 1985; Orlandi et al, 1989), or from mouse spleen (Sastry et al, 1989) from which cDNA for the appropriate gene fragment can be synthesised, cloned and sequenced using conventional methodology.

8.4.2.2   VECTORS

Only a relatively small proportion of cells brought into contact with foreign DNA will become stably transfected with the gene/s. In order to aid selection of the desired cells post-transfection, the vector containing the immunoglobulin DNA should contain markers through which this selection can be made. The vectors of choice are usually of the pSV2 plasmid family, including the SV40 early promoter, providing splice signals and a polyadenylation site. Inserted into the vector are bacterial genes coding for

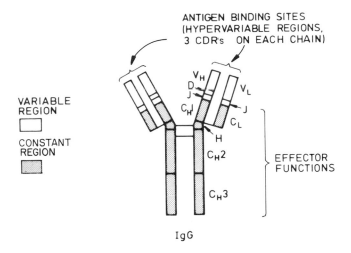

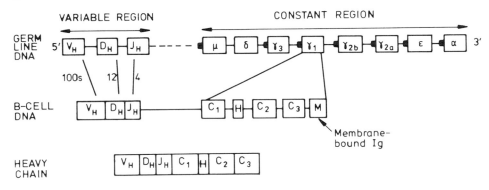

Each constant region gene (except d) has an identical switching sequence at its 5′ end (■), each of which can be selected to combine with the variable region gene sequence. Each constant region gene also contains separate domains which give rise to the characteristic effect or functions of different isotypes

Fig. 8.4.1 Antibody structure and gene sequences

enzymes that block or bypass vital pathways through which metabolic inhibitors can act. The most common selection genes used in immunoglobulin gene transfection are thymidine kinase (tk), xanthine–guanine phosphoribosyl transferase (XGPRT or gpt) and Tn5 transposon phosphotransferase (neo). Cells containing the transfected tk gene can be selected in HAT medium, if the original cell line is deficient in HGPRT. HAT selection can also be used for gpt transfected cells (if the original cell line is deficient in HGPRT) since these cells will express XGPRT (analogous to HGPRT). Alternatively, when this pathway is blocked by mycophenolic acid, cells expressing XGPRT can metabolise exogenous xanthine whereas the endogenous mammalian hypoxanthine cannot. Thus, only the transfected cells will be able to survive by using added xanthine. The neo gene product inactivates the antibiotic G418. Normal protein synthesis is blocked by G418 through its effect on 80S ribosomes. Therefore, only transfected cells will survive in the presence of G418. If required, both selection markers could be used simultaneously in transfecting two different DNA fragments.

8.4.2.3 EXPRESSION

Bacteria have been used extensively to express simple recombinant molecules but they are unable to express functionally active antibody due to the lack of the biosynthetic processes necessary for the glycosylation and folding of complex proteins. With this consideration, the obvious recipient for immunoglobulin DNA transfection is the myeloma cell, since it is specialised in the synthesis and expression of large amounts of immunoglobulin. J558L, a mouse plasmacytoma line expressing λ_1 but no heavy chain (Oi et al, 1983), has proved to be the best to date in terms of transfection frequency, with efficiencies of up to 1 : 1000. The murine hybridoma SP2/0 is not so efficient for transfection purposes but it does have the advantage of being a non-Ig secretor (Boulianne et al, 1984).

The process of transfection can be effected by several means. Calcium phosphate precipitation (Chu & Sharp, 1981), a method used to introduce DNA into many cell types, is not so successful with myeloma cells (Oi et al, 1983). A more consistently efficient and less cytotoxic method was introduced by Sompayrac & Danna (1981) using DEAE-dextran. Alternatively, protoplast fusion with PEG has been used with high efficiency (de Saint Vincent et al, 1981), although this method may be unsuitable for large-scale transfection. Plasmids containing the gene/s of interest are released from *E. coli* by lysozyme digestion of the bacterial cell walls and the resulting spheroplasts are fused directly with myeloma cells by PEG. Electroporation (Potter et al, 1984) is another method involving the application of a high-voltage pulse to myeloma cells in suspension with DNA.

8.4.2.4 SELECTION

The transfected cells are cultured for a period in the appropriate selection medium, depending on the vector used, before large-scale propagation in vitro or in mice. The reconstructed antibody is tested for antigen specificity by immunoassay, and for isotype by immunoassay, if appropriate anti-isotype reagents of sufficient specificity are available, or by SDS-PAGE analysis of biosynthetically labelled antibody.

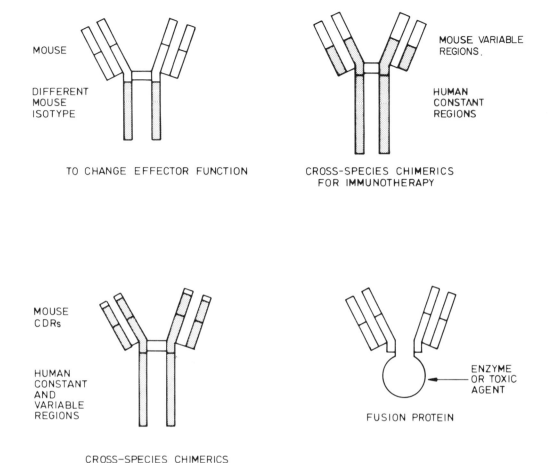

Fig. 8.4.3 Novel antibody constructs

8.4.3 Immunoglobulin gene transfection

In the early 1980s, several groups demonstrated the transfer of immunoglobulin genes into lymphoid cell lines, resulting in the production of intact functional antibody (Rice & Baltimore, 1982; Oi et al, 1983; Neuberger, 1983; Ochi et al, 1983). This achievement opened the way to the production of novel antibodies with different heavy and light chain combinations, different isotypes, cross-species combinations and even non-immunoglobulin effector functions (see Fig. 8.4.3).

8.4.3.1 CROSS-SPECIES COMBINATIONS

Morrison et al (1984) produced chimeric antibodies of mouse antigen-binding domains with human constant region domains. Cloned V_H and V_x genes from an antibody-producing myeloma line were spliced to human C genes ($\gamma1$, $\gamma2$ and x) and inserted into pSV2-gpt and pSV2-neo vectors before transfection into J558L cells and propagation in mice. Boulianne et al (1984) also produced functional mouse–human antibodies of the IgM isotype using similar techniques, with transfection into the Sp2/0 non-producer mouse myeloma cell line. The resulting antibodies retain the antigen specificity of the mouse antibody and the effector function of the human antibody. In therapeutic practice, the adverse immune reaction experienced with mouse antibodies on administration *in vivo* is weaker and delayed with chimeric antibodies (Hale et al, 1983; Giorgi et al, 1983), but some recognition of the mouse variable region remains.

Neuberger et al (1985) constructed an anti-hapten chimeric IgE antibody with a heavy chain composed of human C_ϵ constant region fused to a mouse V_H region. The production of large amounts of human IgE then enabled its use in the study of allergic reactions.

8.4.3.2   CHANGING EFFECTOR FUNCTIONS

It is possible to construct recombinant antibodies with novel non-immunoglobulin effector functions. For example, Neuberger et al (1984), using a mouse IgM gene cloned in the expression vector pSV2gpt, deleted the Fc portion of the heavy-chain gene and replaced it with a mouse γ2b gene containing a DNA restriction fragment of the *S. aureus* nuclease gene. This recombinant gene was transfected into J558L myeloma cells via the plasmid pSV-Vμ_1 and the protein expressed as a F(ab')$_2$-like antibody with enzyme and specific antigen activity. Similarly, Fab-myc proteins incorporating the mouse oncogene c-myc were constructed, suggesting that this approach might be exploited to prepare antisera to previously unidentified products of specific genes.

8.4.3.3   CDR GRAFTING

The specificity with which an antigen binds to its epitope is encoded within the variable regions of the heavy and light chains. When antibodies of different specificities are compared, these regions are shown to contain looped areas in which there is a high degree of variability in amino acid sequence within a conformationally rigid β-sheet framework of relatively constant sequences. It is the hypervariable regions, otherwise known as complementarity-determining regions (CDRs), that are the contact points with the antigen. There are three CDRs on each heavy and light chain.

It is now possible to graft the CDRs from a rodent antibody into the V_H domain of a human antibody, so conferring the specificity of the original rodent antibody on the human antibody (Jones et al, 1986; Verhoeyen et al, 1988; Riechmann et al, 1988). As a result the anti-globulin response during immunotherapy can be further reduced.

The technicalities of these procedures required a knowledge of the amino acid sequences of the variable regions of both antibodies and therefore the nucleotide sequences of the genes (Kabat et al, 1987), so that synthetic oligonucleotides can be made which correspond to the CDRs to be exchanged. Complementary restriction sites are introduced at each end of these oligonucleotides and the framework regions of the recipient human V_H gene so that the original human CDRs can be replaced with those of the rodent antibody. The V_H genes used have been obtained from the human myeloma proteins, NEW and REI, the crystallographic structures of which are known (Saul et al, 1978). Each CDR replacement, for heavy and light chains, is carried out sequentially, each step of which involves transfection into *E. coli*, and plating and probing of colonies with the appropriate primers to distinguish mutant clones from the wild type. Once the complete reshaped variable region genes have been constructed they can be inserted into the appropriate vector together with the desired heavy-chain isotype genes, for transfection into a myeloma cell line as described above.

Small errors in assembling the CDRs need not necessarily seriously affect the binding affinity, since the antigen–antibody complex can adjust slightly to produce the best fit. However, single point mutations at critical positions can substantially increase binding affinity (Allen et al, 1988).

8.4.3.4 SINGLE DOMAIN ANTIBODIES (dAbs)

Natural antibodies and the engineered antibodies described above contain CDR regions from both heavy and light chains which together contribute to the antigen-binding site. However, Ward et al (1989), have shown that V_H domains alone can have binding affinities with antigen comparable with those of the complete antibodies. These authors were able to isolate V_H domains from immune spleen genomic DNA using the polymerase chain reaction, followed by expression in E. coli. These domains can be isolated within days of spleen harvest compared to months of tissue culture to produce conventional monoclonal antibodies. Their relatively small size enables them to penetrate cells with greater facility than complete antibodies and to block otherwise inaccessible epitopes. In future, they may form the basic building block with specificity for antigen, on which antibodies of engineered binding affinity and effector function are based.

Appendix A

Media and buffers

A.1 CELL CULTURE MEDIA AND ADDITIVES

RPMI 1640 wash medium

For all temporary manipulations with cells not involving culture, i.e. centrifuging, washing, diluting etc. Single strength medium can be used directly for this purpose. For reconstitution of the concentrated forms, follow the manufacturers' instructions and the general recommendations in Chapter 2 and Protocol 2.4 in particular.

RPMI 1640 complete culture medium

Preferably prepare only enough for two weeks. Use 10% FCS for most growth conditions involving established cells. Use 20% for fusion, HAT medium and cloning. Established cell lines may be adapted gradually to growth in smaller percentages of FCS, but this must be evaluated for each individual cell line.

To prepare 200 ml final volume:

	10% FCS	20% FCS
RPMI 1640 wash medium (see above)	176 ml	156 ml
Glutamine 200 mM (100× stock)	2 ml	2 ml
Penicillin/streptomycin (100× stock)	2 ml	2 ml
Foetal calf serum*	20 ml	40 ml

Label and store at 4°C.
* FCS should be batch tested for ability to support clonal growth (see Protocol 6.2).

Glutamine 100× stock

Essential nutrient. Working concentration 2 mM. Buy as 200 mM liquid stock or lyophilised powder. Reconstitute with sterile water, aliquot to single use volumes and store at −20°C. Powder stock RPMI medium normally contains glutamine, but unless this is freshly reconstituted, add extra glutamine. Add glutamine to all media kept more than one month at 4°C.

Antibiotics

Penicillin/streptomycin 100× stock

To give basic protection against bacterial infection. The usual working concentration is 100 U/ml penicillin, 100 µg/ml streptomycin. It is most conveniently bought as a mixed 100× concentrated sterile liquid but is also available in powder form. Aliquot aseptically to single use volumes and store at −20°C.

Gentamicin 100× stock

Broader spectrum antibiotic than penicillin/streptomycin. The usual working concentration is 200 µg/ml. Buy as a 100× sterile stock solution. Aliquot aseptically to single use volumes and store at 4°C. Stock solution very stable.

Amphotericin B 100× stock

Offers some protection against yeast and fungal infection. Working concentration is 2.5 µg/ml. Buy as lyophilised powder and reconstitute to 100× stock solution using sterile H_2O. Do not filter, autoclave or heat above 50°C but it can bear repeated freeze–thawing. Do not store for more than one month at 4°C. For prolonged storage, aliquot and store at −20°C. The half-life at 37°C is five days and the toxic level is only 10 µg/ml, so be particularly careful when topping up the dose.

Selection media

HAT medium

Selective medium for hybridomas of HGPRT– myeloma parentage. To prepare 200 ml final volume:

	10% FCS	20% FCS
RPMI 1640 wash medium (see above)	172 ml	152 ml
Glutamine 200 mM (100× stock)	2 ml	2 ml
Penicillin/streptomycin (100× stock)	2 ml	2 ml
Hypoxanthine/thymidine (HT) (100× stock)	2 ml	2 ml
Aminopterin (A) 100× stock	2 ml	2 ml
Foetal calf serum	20 ml	40 ml

Label and store at 4°C.

HT medium

Use as transition medium for HAT-selected hybridomas before transfer to RPMI medium.
Prepare as for HAT medium but replace aminopterin with 2 ml of RPMI wash medium.

HT 100× stock

For HAT and HT medium:
This can be bought as a ready to use solution or alternatively can be made up from powder.

Dissolve 136.1 mg hypoxanthine and 38.8 mg thymidine in 100 ml double distilled water at 50°C. Sterilise by membrane filtration, aliquot and store at −20°C.

For use:
Add 1 ml to 100 ml medium. Final concentration 1×10^{-4} M hypoxanthine, 1.6×10^{-5} M thymidine.

Aminopterin 100× stock

For HAT medium:
This can also be bought as a ready to use solution and in view of its potential toxicity is best bought in this form. If it is to made made up from powder, wear a mask and gloves.

Add 1.76 mg aminopterin to 90 ml double distilled water. Add 1 M NaOH dropwise until the aminopterin dissolves. Titrate to pH 7.5 with 1 M HCl and adjust to 100 ml with water. Sterilise by membrane filtration and store in aliquots at −20°C.

Culture media buffers

NaHCO₃ 7.5% stock

Buffer to be added to powder media:
Either prepare a 7.5 g/100 ml solution (add 27 ml per litre of RPMI medium) or add 2 g dry powder to one litre RPMI medium. Filter-sterilise.

HEPES

Either buy the Dutch modification of RPMI powder which will contain the appropriate concentrations of HEPES and NaHCO₃ or add 20 mM HEPES (4.77 g/litre of final medium) and 17 ml of 5% NaHCO₃ (0.85 g/litre) to RPMI powder without NaHCO₃.

Optional supplements

2-Mercaptoethanol 100× stock

Mix in a fume hood 0.5 ml of commercial stock with 6.6 ml water (i.e. 1 M).
Mix 5 ml of 1 M solution with 95 ml water (5×10^{-2} M).
Mix 10 ml of 5×10^{-2} M with 90 ml basic salt solution or RPMI medium (5×10^{-3} M).
Filter-sterilise and aliquot.
Store at 4°C.
For use add 1 ml 5×10^{-3} M to 100 ml growth medium.

Sodium pyruvate

Working concentration 1 mM. Buy 100 mM stock and store aliquoted in single use volumes at −20°C for four to six months.

Miscellaneous reagents for cell culture

8-Azaguanine or 6-thioguanine

For routine 'clearing' of HGPRT⁺ revertant myeloma cells:
Add 20 μg/ml to growth medium for a few growth cycles, at monthly intervals. Human
   myeloma cells may need a higher and constant dose.
Dissolve 120 mg azaguanine or thioguanine in 10–20 ml 0.1 M NaOH.
Make up to 40 ml with water, i.e. 3 mg/ml.
Filter, aliquot and store at −20°C.
For use add 660 μl to 100 ml medium.

PEG 1500, 40 or 50%

For fusion:
This can be bought as a sterile liquid ready for use or a 40% or 50% solution can be made up from the waxy solid in physiological saline. Batches will vary due to varying levels of impurities. If not infected, the stock will have a shelf life of at least a year.

Trypsin

For removal of firmly attached cells from culture vessel (scraping is adequate for most myeloma and hybridoma cells):

Wash cells in Ca^{2+}/Mg^{2+}-free Hanks solution (from tissue culture medium suppliers).
Add 1% trypsin in 0.02% EDTA for approximately 8 min.
Add equal volume of RPMI/10% FCS.
Remove cells and wash twice by centrifuging at 400 g and resuspending cells in fresh
   culture medium.
Distribute as required.

Freezing medium for cells

Recipes vary slightly. Generally, they are FCS or RPMI/20% FCS containing 4–10% DMSO (dimethyl sulphoxide). Prepare before use but test the freezing procedure and recovery on myelomas before risking precious hybrids.

Cell viability stains

(1)   Ethidium bromide/acridine orange

Dissolve 50 mg ethidium bromide and 15 mg acridine orange in 1 ml ethanol.
Add 49 ml water.
Mix and aliquot (1 ml) and store at $-20°C$.
For use dilute 1 ml aliquot with 99 ml PBS, mix and store in brown bottle for up to
   one month.
Mix equal volumes of stain and cell sample and count immediately under fluorescent
   microscope.

(2)   0.2% Nigrosin dye

Prepare a 2% stock solution of nigrosin, dilute further with sterile saline for use.

A.2   ASSAY BUFFERS

Coating buffers (i.e. ELISA and IRMA)

For coating antibody or antigen to the solid phase:

(1)   Carbonate/bicarbonate buffer, pH 9.6

   1.59 g Na_2CO_3
   2.93 g $NaHCO_3$
   0.2 g Thiomersal (optional)
   Make up to one litre and store at 4°C for no more than two weeks.

(2)   10 mM Tris-HCl, pH 8.5 + 100 mM NaCl

   1.12 g Tris
   5.8 g NaCl
   900 ml H_2O
   Adjust to pH 8.5 and make up to one litre.

(3) 10 mM sodium phosphate, pH 7.2 + 100 mM NaCl

     1.78 g/l $Na_2HPO_4.2H_2O$
     1.56 g/l $NaH_2PO_4.2H_2O$
     Mix 360 ml Na_2HPO_4 with 140 ml NaH_2PO_4.
     Add 5.8 g NaCl and make up to one litre.

(4) Poly-L-lysine for coating cells or small synthetic peptides

     Dissolve poly-L-lysine (1 mg/ml, 0.1%) in required volume of PBS.
     Incubate 50 μl/well for 90 min at 37°C prior to adding cells.

Wash buffers

(1) PBS-Tween (0.05%)

NaCl	8.0 g
KH_2PO_4	0.2 g
$Na_2HPO_4.12H_2O$	2.9 g
KCl	0.2 g
Tween 20	0.5 ml

     Make up to one litre, pH 7.4. Store at 4°C.

(2) 3 M Tris-HCl, pH 8.0
     For assays using alkaline phosphatase:

     Dissolve 181.65 g Tris in 450 ml distilled water.
     Titrate to pH 8 with 1 M HCl.
     Adjust to 500 ml with distilled water.

Block solutions

1% BSA, gelatin, ovalbumin, casein, milk powder.

Enzyme substrates

For horseradish peroxidase (HRP)

(1) Orthophenylene diamine (OPD)
     (in 0.1 M citrate-phosphate buffer, pH 5.0)

     Prepare stock solutions of 0.1 m citric acid and 0.2 M phosphate:
       19.2 g/l citric acid
       28.4 g/l Na_2HPO_4
     Store at 4°C

     Mix 24.3 ml 0.1 M citric acid +
         25.7 ml 0.2 M Na_2HPO_4 +
         50 ml H_2O
     Immediately before use add 40 mg OPD and 40 μl of 30% H_2O_2.
     Stop colour development with 2.5 M H_2SO_4 (i.e. 50 μl/well) and read at 492 nm.

(2) 5-Aminosalicylic acid (5-AS)

   Prepare purified stock by dissolving 5 g 5-AS and 5 g sodium bisulphite in 550 ml
      deionized water at 80°C.
   Maintain temperature with stirring for 10 min.
   Add 2 g activated charcoal and mix for 5 min.
   Filter and cool to 4°C.
   Wash precipitate twice with 5 ml water.
   Store desiccated in dark or redissolve 1 g/litre of 10 mM phosphate buffer, pH 6.0,
      containing 0.1 mM EDTA.
   Store aliquoted at $-20°C$.
   For use, redissolve precipitate by warming to room temperature.
   Add H_2O_2 (0.003%) just before use.
   Stop colour development with 0.01% w/v sodium azide in 0.1 M citric acid and
      read at 492 nm.

(3) 2,2′-Azino-di-(3-ethylbenzthiazoline-6-sulphonic acid) (ABTS)

   Dissolve 40 mg ABTS in 100 ml of 100 mM phosphate-citrate buffer, pH 4.0,
      containing 0.002% H_2O_2.
   Stop colour development with 0.01% w/v sodium azide in 0.1 M citric
      acid.
   The blue–green product absorbs at 415 nm.

(4) 3,3′,5,5′-Tetramethylbenzidine (TMB)

   Dissolve 5 mg TMB in 2.5 ml absolute ethanol (heat to 40°C if necessary).
   Prepare 200 mM acetate buffer, pH 3.3, and add to 92.5 ml distilled water containing
      100 mg nitroferricyanide.
   Mix the two solutions and add 0.003% H_2O_2 just before use.
   Stop colour development with 2 M H_2SO_4.
   The blue or yellow product absorbs at 655 or 450 nm.

For alkaline phosphatase (AP)

(1) *p*-Nitrophenyl phosphate (p-NPP)

   Mix 0.01% $MgCl_2$ (10 mg/100 ml) p-NPP (100 mg/100 ml) in 100 mM Tris-HCl,
      pH 8.1.
   Stop colour development with 2 M NaOH and read the pale yellow product
      at 405 nm.

For β-galactosidase (β-G)

(1) *o*-Nitrophenyl-β-galactosidase (ONPG)

   Dissolve 0.9 mg/ml ONPG in PBS, pH 7.5 containing 10 mM $MgCl_2$ (0.2 g/100 ml)
      and 0.1 M 2-ME (700 μl).
   Stop the reaction with 1 M Na_2CO_3. The yellow product absorbs at 410 nm.

For urease

(1) Bromocresol purple (BCP)

Dissolve 8 mg BCP in minimal volume 0.01 M NaOH.
Make up to 100 ml in deionised water.
Add 100 mg urea and EDTA to a final concentration of 0.2 mM.
Adjust pH to 4.8 with 0.1 M NaOH or 0.1 M HCl and store at 4°C.

Immunocytochemistry and immunoblotting

Enzyme substrates

For horseradish peroxidase

(1) Diaminobenzidine (DAB)

Dissolve 0.5 mg/ml DAB in Tris- HCl or PBS, pH 7.4, containing 0.02% H_2O_2.
Insoluble brown product.
Can be enhanced by adding cobalt or nickel chloride to 0.03% w/v.

(2) 4-Chloro-1-naphthol (4-CN)

Dissolve 40 mg 4-CN in 0.5 ml of absolute ethanol. While stirring, add to 100 ml
of 50 mM Tris-HCl buffer or PBS, pH 7.6, containing 0.03% H_2O_2, preferably at
50°C. Filter and use immediately.
Insoluble blue–black product.

(3) 3-Amino-9-ethylcarbazole (AEC)

Prepare 0.5 ml of 0.4% AEC in DMF in 9.5 ml 50 mM acetate buffer, pH 5,
    containing 1–10 μl 30% H_2O_2.
Filter the solution onto sections.
Leave for 3–10 min at room temperature.
Insoluble red product.

(4) Hanker–Yates reagent

Place 75 mg HY reagent to each of six chemically clean and dry Coplin jars.
Just prior to immersing slides, add 200 mM Tris-HCl, pH 7.6, and 20 μl of 30%
    H_2O_2.
Agitate solution during colour development. When solution turns brown move
    slide to fresh Coplin jar.
Purple–brown, almost black, product. Does not react with endogenous peroxidase.

For alkaline phosphatase

(1) Bromochloroindolyl phosphate (BCIP) and nitroblue tetrazolium (NBT)

Dissolve 50 mg/ml of BCIP in 100% dimethylformamide (DMF) and 75 mg/ml of
    NBT in 70% DMF.
Mix 33 μl of BCIP stock, 25 μl of NBT stock and 7 ml of 100 mM Tris-HCl buffer,
    pH 9.0, containing 0.01% $MgCl_2$.
A blue precipitate is formed.

(2) Fast Red/Blue

Prepare stock solution of 10 mg/ml of naphthol AS-MX, or naphthol AS phosphate or naphthol AS-BI phosphoric acid sodium salt in DMF, diluted 50 times with 100 mM Tris-HCl buffer, pH 8.2.

Store at 4°C for up to several weeks.

Before use dissolve 1 mg/ml Fast Red TR or Fast Blue BBN in the naphthol stock solution.

For glucose oxidase

(1) Glucose, T-nitroblue tetrazolium chloride (t-NBT) and M-phenazine methosulphate (m-PMS)

Dissolve 2.5 mM t-NBT and 0.6 mM m-PMS in 100 mM phosphate buffer, pH 6.6, containing 41.7 mM glucose, approximately one hour before staining.

After application of stain, place slides in opaque plastic box, close and immerse in 50°C waterbath for 30 min.

Wash several times with PBS before mounting.

Mountants for FITC staining

(1) DABCO

Dissolve 0.25 g DABCO (Sigma D-2522) in 9 ml of warm glycerol + 1 ml PBS.

(2) Moviol Mountant for Hoechst stain

6 g (~4.7 ml) glycerol (analytical grade)
2.4 g Moviol 4-88 (Hoechst)
Mix thoroughly in 50-ml plastic conical centrifuge tube. Add 6 ml distilled H_2O and leave for two hours at room temperature. Add 12 ml 0.2 M Tris-HCl buffer, pH 8.5, and incubate in 70°C waterbath for 10 min with occasional stirring. Centrifuge at 5000 g for 15 min and aliquot in glass vials.

Miscellaneous buffers

Refer also to *Data for Biochemical Research* (1986, 3rd Edition), Eds Dawson, Elliott, Elliott & Jones, Oxford Science Publications.

Alsevers solution (for storing red blood cells)

1.6 g trisodium citrate
0.84 g NaCl
4.1 g glucose
Mix in 150 ml distilled H_2O, bring pH to 6.1 with 10% citric acid solution, and make up to 200 ml.

0.1 M Borate buffer, pH 8.5

Disodium tetraborate ($Na_2B_4O_7$. 10 H_2O) 9.54 g in 250 ml distilled water
Boric acid 24.73 g in four litres distilled water.
Add borate solution to the four litres of boric acid solution until the pH reaches 8.5.

Borate saline buffer, pH 8.3

Boric acid, 6.18 g/litre
Sodium tetraborate (borax) 9.54 g/litre
Sodium chloride, 4.38 g/litre
Make up to one litre with distilled water.

0.2 M Carbonate buffer, pH 9.5

Sodium carbonate, Na_2CO_3, 21.2 g/litre
Sodium bicarbonate, $NaHCO_3$, 16.8 g/litre
Add sodium carbonate solution to sodium bicarbonate solution to pH 9.5 (approx.
    6.4 ml to 18.6 ml).

Con A buffer

100 mM acetate, pH 6.0, + 1 M NaCl, 1 mM $CaCl_2$, 1 mM $MgCl_2$, 1 mM $MnCl_2$
EDTA (for use as anticoagulant)

Prepare EDTA solution of 100 g/litre.
Effective anticoagulant at 1.5 ± 0.25 mg/ml of blood.
Allow appropriate volume to dry in blood collection bottles at 20°C.
Mix thoroughly after blood collection.

PBS (0.01 M phosphate/0.15 M NaCl, pH 7.2)

One litre of 10× stock solution:
NaCl 80 g
KCl 2 g
Na_2HPO_4 (anhydrous) 11.5 g
KH_2PO_4 2 g
Store at 4°C and add 0.2 g thiomersal as preservative if desired. Dilute 1 : 10 prior to use.

Saturated ammonium sulphate (SAS)

1000 g ammonium sulphate in one litre of H_2O at 50°C.
Stand overnight at room temperature.
Adjust pH to 7.2 with dilute ammonia or H_2SO_4.

0.02 M Tris-NaCl, pH 7.2 (Affi-Gel column buffer)

2.42 g Tris
25 mM NaCl (1.45 g/litre)
900 ml H_2O
Adjust pH to 7.2 with 1 M HCl. Make up to one litre.
Add 50 mM NaCl (2.9 g/litre) to above buffer for Ig elution.

Appendix B

Sources of equipment and reagents

The following list is not meant to be exhaustive, merely a guide to the main suppliers in the UK and USA. Some items are known by the manufacturer's name and are marketed by other companies. Where possible both names are given and the addresses can be found in Appendix C. The list is not intended to be a recommendation or endorsement of individual products by the authors.

Item	Suppliers
Anemometers	3
Antibiotics	9, 32, 36, 42, 61, 74
Autoclaves/ovens	25, 38, 44
Bolton–Hunter reagent	6
Celite	18
Cell lines	5, 29
Centrifuges	24, 33, 56, 70
Clean rooms	32
CO_2 cylinders	15
Cylinder change over device	58
Cylinder valves	15

Appendix C

Addresses of suppliers

1. Accurate Chemical & Scientific Corp., 300 Shames Drive, Westbury NY 11590. Tel. 800 645 6264
2. Advanced Protein Products Ltd, Unit 18H, Premier Partnership Estate, Leys Road, Brockmoor, Brierley Hill, West Midlands DY5 3UP. Tel. 0384 263862
3. Airflow Developments Ltd, Lancaster Rd, High Wycombe, Bucks HP12 3QP. Tel. 0494 25252
4. Alpha Laboratories Ltd, 40 Parham Drive, Eastleigh, Hants SO6 4NU. Tel. 0703 610911
5. American Type Culture Collection (ATCC), 12301 Parklawn Drive, Rockville, MD 20852. Tel. 800 638 6597; 301 231 5594
6. Amersham International PLC, White Lion Road, Amersham, Bucks HP7 9LL. Tel. 02404 4444
   Amersham Corporation, 2636 South Clearbrook Drive, Arlington Heights, IL 60005. Tel. 800 323 9750
7. Anachem Ltd, Anachem House, Charles St, Luton, Beds LU2 0EB. Tel. 0582 456666
8. Avon Medical, Moons Moat North, Redditch, Worcs B98 9HA. Tel. 0527 64901
9. BCL Ltd, Boehringer Mannheim House, Bell Lane, Lewes, East Sussex BN7 1LG. Tel. 0273 480444
   Boehringer Mannheim Biochemicals, P.O. Box 50816, Indianapolis, IN 46250. Tel. 800 262 1640
10. BDH Ltd, Broom Rd, Poole, Dorset BH12 4NN. Tel. 0202 745520

11. Beckman RIIC Ltd, Progress Rd, Sands Industrial Estate, High Wycombe, Bucks HP12 4JL. Tel. 0494 41181
    Beckman Instruments Inc., 2470 Faraday St, Carlsbad, CA 92008. Tel. 619 438 9151
12. Beckton-Dickinson UK Ltd, Between Towns Road, Cowley, Oxford OX4 3LY
    Becton-Dickinson Immunocytometry Systems, P.O. Box 7375, Mountain View, CA 94309. Tel. 800 223 8226
    Denderstraat 24, B-9440, Erembodegem, Belgium. Tel. 053 787830
13. Biological Industries Ltd, 56 Telford Rd, Cumbernauld, Glasgow G67 2AX. Tel. 02367 28700
14. Bio-Rad Laboratories Ltd, Caxton Way, Watford Business Park, Watford, Hertfordshire WD1 8RP. Tel. 0923 40322
    Bio-Rad Laboratories, 1414, Harbour Way South, Richmond, CA 94804. Tel. 800 424 6723
15. BOC Ltd, (local distributors) P.O. Box 12, Priestley Rd, Worsley, Manchester M28 4UT. Tel. 061 794 4651
16. Boehringer Mannheim Biochemicals (see BCL)
17. CABS (Current Awareness in Biological Sciences), Subscriptions: Pergamon Press plc, Headington Hill Hall, Oxford, OX3 0BW
    Pergamon Press Inc., Maxwell House, Fairview Park, Elmsford, NY 10523
18. Calbiochem, P.O. Box 12087, San Diego, CA 92112 4180. Tel. 800 854 9256
    Novabiochem (UK) Ltd, 3 Heathcote Building, Highfields Science Park, University Boulevard, Nottingham NG7 1BR. Tel. 0602 430840
19. Cambio Ltd, 34 Millington Rd, Cambridge CB3 9HP. Tel. 0223 66500
20. Cambridge Bioscience, 42 Devonshire Rd, Cambridge CB1 2BL. Tel. 0223 316855
21. Cambridge Research Biochemicals Ltd, Gadbrook Park, Northwich, Cheshire CW9 7RA. Tel. 0606 41100
22. Ciba Corning Diagnostics Ltd, Colchester Rd, Halstead, Essex CO9 2DX. Tel. 4478 747 4742
    Ciba Corning Diagnostics, 63 North St, Medfield, MA 02052. Tel. 800 255 3232
23. Dakopatts Ltd, 22 The Arcade, The Octagon, High Wycombe, Bucks HP11 2HT. Tel. 0494 452016
    Dako Corporation, 22 North Milpas St, Santa Barbara CA 93103. Tel. 800 235 5743
24. Damon/IEC (UK) Ltd, Unit 7, Lawrence Way, Brewers Hill Rd, Dunstable, Beds LU6 1BD. Tel. 0582 604669
    Damon Biotech Inc., 119 Fourth Ave., Needham Heights, MA 02194. Tel. 617-449-6002
25. Denley Instruments Ltd, Natts Lane, Billingshurst, Sussex RH14 9EY. Tel. 040381 3441
26. Difco Laboratories Ltd, P.O. Box 14b Central Ave., East Moseley KT8 0SE. Tel. 01 979 9951
    Difco Labs, P.O. Box 1058, Detroit, MI 48232. Tel. 313 961 0800
27. Dynal A/S, P.O. Box 158 Skoyen, N-0212 Oslo 2, Norway. Tel. 472 507800
    Dynal (UK) Ltd, Station House, 26, Grove St, New Ferry, Wirral, Merseyside L62 5AZ. Tel. 051 644 6555
    Dynal Inc., 45 North Station Plaza, 406 Great Neck, NY 11021. Tel. 516 829 0039
28. Dynatech Laboratories Ltd, Daux Rd, Billingshurst, Sussex RH14 9SJ. Tel. 040381 3381
    Dynatech Inc., 14340, Sullyfield Circle, Chantilly, VA 22021. Tel. 703 631 7800
29. European Collection of Animal Cell Cultures (ECACC), PHLS CAMR, Porton Down, Salisbury, Wiltshire. Tel. 0890 619391

30. Endotronics UK (see Technogen Systems Ltd)
    Endotronics Inc., 8500 Evergreen Boulevard, Coon Rapids, MN 55433. Tel. 612 786 0302
31. Envair (UK) Ltd, Bank Chambers, Ratcliffe St, Haslingdon, Rossendale, Lancs BB4 5DE. Tel. 0706 227931
32. Flow Laboratories Ltd, Woodcock Hill, Harefield Road, Rickmansworth, Hertfordshire WD3 1PQ. Tel. 0923 774 666
    Flow Laboratories Inc., 7655 Old Springhouse Rd, McLean, VA. Tel. 800 368 3569
33. Gallenkamp, Belton Rd West, Loughborough LE11 0TR. Tel. 0509 237371
34. GCA Corporation, Precision Scientific Group, 3737 West Cortland St, Chicago, IL 60647. Tel. 312 227 2660
35. Gelman Sciences Ltd, Brackmills Business Park, Caswell Rd, Northampton NM4 0EZ. Tel. 0604 765141
    Gelman Inc., 600 South Wagner Rd, Ann Arbor, MI 48106. Tel. 800 521 1520
36. GIBCO Ltd, (now Life Technologies Ltd) P.O. Box 35, Trident House, Renfrew Road, Paisley, Scotland PA3 4EF. Tel. 041 889 6100
    GIBCO, 3175 Staley Rd, Grand Island, NY 14072. Tel. 800 828-6686
37. Harlan Sprague Dawley Inc., P.O. Box 29176, Indianapolis, IN 46229. Tel. 317 894 7521
38. Heraeus Equipment Ltd, Unit 9, Wates Way, Ongar Rd, Brentwood, Essex. Tel. 0277 231511
    Heraeus Inc. Equipment Group, 111A Corporate Boulevard, South Plainfield, NJ 07080
39. Home Office, 50 Queen Anne's Gate, London SW1H 9YN
40. ICI Pharmaceuticals Division, P.O. Box 35, Macclesfield, Cheshire. Tel. 0625 512023
    ICI Pharmaceutical Group, ICI America Inc., P.O. Box 751, Wilmington, DE 19897
41. ICN Biomedical Ltd, Free Press House, Castle Street, High Wycombe, Bucks HP13 6RN. Tel. 0494 443826
    ICN Biomedicals Inc.,3300 Hylands Ave., Costa Mesa, CA 92626. Tel. 714 545 0113
    ICN Immunobiologicals, 4720 Yender Ave., P.O. Box 1200, Lisle, IL 60532. Tel. 800 348 7465
42. Imperial Laboratories, West Portway, Andover, Hants SP10 3LF. Tel. 0264 333311
43. Janssen Life Sciences (also from Amersham), Grove Wantage, Oxon OX12 0DQ. Tel. 02357 2966
    Janssen Inc., 40 Kingsbridge Rd, Piscataway, NJ 08854. Tel. 201 524 0400
44. Jencons (Scientific) Ltd, Cherrycourt Way Industrial Estate, Stanbridge Road, Leighton Buzzard, Bedfordshire LU7 8UA
45. Labsystems (UK) Ltd, 12 Redford Way, Uxbridge, Middlesex UB8 1SZ. Tel. 0895 38421
46. Laboratorium Professor Berthold, 07547 Wildbad, Germany
47. Laboratory Impex Ltd, 111–113 Waldegrave Road, Teddington, Middlesex TW11 8LL. Tel. 4401 977 3266
48. LH Fermentation, Porton House, Vanwall Rd, Maidenhead SL6 4UB. Tel. 0628 771471
49. Life Science Labs Ltd, Sedgewick Rd, Luton, Beds LU4 9DT. Tel. 0582 597676
50. Linscotts Directory, P.O. Box 55, East Grinstead, Sussex RH19 3YL. Tel. 0342 824854
    40 Glenn Drive, Mill Valley, CA 94941. Tel. 415-383-2666
51. LKB (see Pharmacia)
52. M.D.H. Ltd, Walworth Rd, Andover, Hampshire SP10 5AA

53. Microgon Inc. (UK; see New Brunswick) 23152 Verdugo Drive, Laguna Hills, CA 92653. Tel. 714-581-3880
54. Miles (see ICN)
55. Millipore (UK) Ltd, The Boulevard, Blackmoor Lane, Watford, Herts WD1 8WD. Tel. 0923 816375
    Millipore Corp., 80 Ashby Rd, Bedford, MA 01730. Tel. 617 275 9200
56. MSE, Bishop Meadow Rd, Loughborough, Leicester LE11 0RG. Tel. 0509 616000
57. Nalgene Co. (UK Distributor, see BI), P.O. Box 20365, Rochester, NY 14602-0365. Tel. 716 586 8800
58. New Brunswick Scientific (UK) Ltd, Edison House, Unit 7, A1M Business Centre, 163 Dixons Hill Rd, North Mimms, Herts AL9 7JE. Tel. 0707 275733
    New Brunswick Scientific Co Inc., P.O. Box 987, 44 Talmadge Rd, Edison, NJ 08818-4005. Tel. 201 287 1200
59. Nikon UK Ltd, Haybrook, Halesfield 9, Telford, Shropshire IF7 4EW. Tel. 0952 587 444
    Nikon Inc., 623 Stewart Ave., Garden City, NY 11530. Tel. 516 222 0200
60. Nordic Immunological Labs Ltd, Dairy House, Moneyrow Green, Holyport, Maidenhead, Berks. Tel. 0628 24978
    Nordic Immunological Labs, Drawer 2517, Capistrano Beach, CA 92624. Tel. 800 554 6655
61. Northumbria Biologicals Ltd, South Nelson Industrial Estate, Cramlington, Northumberland NE23 9HL
62. Nycomed UK Ltd, Nycomed House, 2111 Coventry Rd, Sheldon, Birmingham B26 3EA (USA Distributor, see ACSC)
63. Nyegaard UK Ltd, Nyegaard House, 2111 Coventry Rd, Sheldon, Birmingham B26 3AE. Tel. 021 742 8806
64. OLAC Ltd, Shaw's Farm, Blackthorn, Bicester, Oxon OX6 0TP. Tel. 0869 243241
65. Olympus Optical Co. Ltd, 2–8 Honduras St, London EC1 YOTX
    Olympus Corporation, 4 Nevada Drive, Lake Success, NY 11042-1179
66. Pharmacia Ltd, Pharmacia House, Midsummer Boulevard, Central Milton Keynes, Bucks MK9 3HP
    Pharmacia LKB Biotechnology Inc., 800 Centennial Ave., P.O. Box 1327, Piscataway, NJ 08855-1327. Tel. 201 457 8000
67. Polysciences Inc., 400 Valley Rd, Warrington, PA 18976. Tel. 215 343 6484
68. Richardsons of Leicester Ltd, Evington Valley Rd, Leicester LE5 5LJ. Tel. 0533 736571
69. Sabre, Manor Farm Rd, Reading RG2 0LG. Tel. 0734 876111
70. Sarstedt Ltd, 68 Boston Rd, Beaumont Leys, Leicester LE4 1AW. Tel. 0533 359023
    Sarstedt Inc., P.O. Box 468, Newton, NC 28658/0468. Tel. 704 465 4000
71. Sartorius Instruments Ltd, Longmead, Blenheim Rd, Epsom, Surrey KT19 9QN. Tel. 01 642 8691
72. Serotec Ltd, 22 Bankside, Station Approach, Kidlington, Oxford OX5 1BR. Tel. 08675 79941
    US Distributor: Biproducts for Science Inc., P.O. Box 29176, Indianapolis, IN 46229. Tel. 317 894 7536
73. Serva Fine Biochemicals (UK Distributor, see Cambridge Bioscience), 200 Shames Drive, Westbury, NY 11590. Tel. 800 645 3412
74. Sigma Chemical Co. Ltd, Fancy Rd, Poole, Dorset BH17 7NH. Tel. 0800 373731
    Sigma Chemical Co., P.O. Box 14508, St Louis, MO 63178. Tel. (800) 325-3010
75. Sterilin Ltd, Lampton House, Lampton Road, Hounslow, Middlesex TW3 4EE. Tel. 01 572 2468

76. SUBIS, University, Sheffield S10 2TN. Tel. 0742 768555 Ext. 6232
77. Technicon Instrumentation Company Ltd, Evans House, Houndsmill Industrial Estate, Basingstoke, Herts. Tel. 0256 29181
    Technicon Instrumentation Corporation, 511 Benedict Ave., Tarry Town, NY 10591. Tel. 914 631 8000
78. Technogen Systems Ltd, Altec House, Brigade Close, South Harrow, Middlesex HA2 0LQ. Tel. 01 423 6333
79. University Federation for Animal Welfare (UFAW), 8 Hamilton Close, South Mimms, Potters Bar, Herts EN6 3QD
80. Wellcome Diagnostics Ltd, Temple Hill, Dartford DA1 5A8. Tel. 0322 77711
    Wellcome Diagnostics Division, North Building, 3030 Cornwallis Rd, Research Triangle Park, NC 27709. Tel. 800 334 9332
81. Carl Zeiss, D-7082 Oberkochen, Germany
    P.O. Box 78, Woodfield Rd, Welwyn Garden City, Herts AL7 1LU. Tel. 0707 331144
    1 Zeiss Drive, Thornwood, NY 10594. Tel. 914 681 7752

References

Aarden, L., Lansdorp, P. and de Groot, E. (1985) A growth factor for B cell hybridomas produced by human monocytes. *Lymphokines* **10**: 175

Abrams, P. G., Knost, J. A., Clarke, G., Wilburn, S., Oldham, R. K. and Foon, K. A. (1983) Determination of the optimal human cell lines for development of human hybridomas. *J. Immunol.* **131**: 1201–1204

Abrams, P. G., Rossio, J. L., Stevenson, H. C. and Foon, K. A. (1986) Optimal strategies for developing human–human monoclonal antibodies. *Methods Enzymol.* **121**: 107–119

Adams, J. C. (1981) Heavy metal intensification of DAB-based HRP reaction product. *J. Histochem. Cytochem.* **29**: 775

Allen, D., Cumano, A., Simon, T., Sablitzky, F. and Rajewsky, K. (1988) Modulation of antibody binding affinity by somatic mutation. *Int. J. Cancer. Suppl.* **3**: 1–8

Anderson, D. V., Clarke, S. W., Stein, J. M. and Tucker, E. M. (1986) Bovine and ovine monoclonal antibodies to erythrocyte determinants, produced by interspecies hetero-myelomas. *Biochem. Soc. Trans.* **14**: 72

Anderson, P. N. and Potter, M. (1969) Induction of plasma cell tumours in Balb/c mice with 2, 6, 10, 14-tetramethylpentadecane (Pristane). *Nature* **222**: 994–995

Astaldi, G. C. B., Jansson, M. C., Lansdorp, P., Williams, C., Zeijlemaker, W. P. and Oosterhof, F. (1980) Human endothelial cell culture supernatant (HECS). A growth factor for hybridomas. *J. Immunol.* **125**: 1411–1414

Avrameas, S. (1969) Coupling of enzymes to proteins with glutaraldehyde. Use of the conjugates for the detection of antigens and antibodies. *Immunochemistry* **6**: 43–52

Avrameas, S. and Ternynck, T. (1969) The cross-linking of proteins with glutaraldehyde and its use for the preparation of immunoadsorbents. *Immunochemistry* **6**: 53–66

Avrameas, S. and Ternynck, T. (1971) Peroxidase labelled antibody and Fab conjugates with enhanced intracellular penetration. *Immunochemistry* **8**: 1175

Bardsley, D. W., Coakley, W. T., Jones, G. and Liddell, J. E. (1989) Electroacoustic fusion of millilitre volumes of cells in physiological medium. *J. Biochem. Biophys. Methods* **19**: 339–348

Bardsley, D. W., Liddell, J. E., Coakley, W. T. and Clarke, D. J. (1990) Electroacoustic fusion of murine hybridomas. *J. Immunol. Methods* **129**: 41–47

Bazin, H. (1982) Production of rat monoclonal antibodies with the Lou rat non-secreting IR983F myeloma cell line. In *Protides of the Biological Fluids* (H. Peeters, ed.), Pergamon Press, Oxford, pp. 615–618

Bazin, R. and Lemieux, R. (1987) Role of the macrophage-derived hybridoma growth factor in the in vitro and in vivo proliferation of newly formed B cell hybridomas. *J. Immunol.* **139**: 780

Bazin, R. and Lemieux, R. (1989) Increased proportion of B cell hybridomas secreting monoclonal antibodies of desired specificity in cultures containing macrophage derived hybridoma growth factor (IL-6). *J. Immunol. Methods* **116**: 245–249

Berg, H. (1982) Molecular biological implications of electric-field effects. *Studia Biophysica* **90**: 169–176

Birch, J. R., Boraston, R. and Wood, L. (1985) Bulk production of monoclonal antibodies in fermenters. *Trends Biotechnol.* **3**: 162–166

Bischoff, R., Eisert, R. M., Schedel, I., Vienken, J. and Zimmermann, U. (1982) Human hybridoma cells produced by electrofusion. *FEBS Lett.* **147** (1): 64–67

Bolton, A. E. and Hunter, W. M. (1973) The labelling of proteins to high specific radioactivities by conjugation to a ^{125}I-containing acylating agent. *Biochem. J.* **133**: 529–539

Borrebaeck, C. A. K. (ed.) (1988) *In Vitro Immunisation in Hybridoma Technology*. Progress in Biotechnology Vol. 5. Elsevier Science Publishers, Amsterdam.

Borrebaeck, C. A. K. & Hagen, I. (eds) (1989) Electromanipulation in Hybridoma Technology: A Laboratory Manual. Macmillan: Stockton Press

Bos, E., van der Doelen, A. A., van Rooy, N. and Schuurs, A. H. (1981) 3,3′,5,5′-Tetramethylbenzidine as an Ames test negative chromogen for horseradish peroxidase in enzyme-immunoassay. *J. Immunoassay* **2**: 187–204

Boss, B. (1984) An improved in vitro immunisation procedure for the production of monoclonal antibodies against neural and other antigens. *Brain Res.* **291**: 193–196

Boulianne, G. L., Hozumi, N. and Schulman, M. J. (1984) Production of functional mouse/human antibody. *Nature* **312**: 643–646

Boyden, S. V. (1951) The adsorption of proteins on erythrocytes treated with tannic acid and subsequent hemagglutination by anti-protein sera. *J. Exp. Med.* **93**: 107

Boyum, A. (1968) Isolation of mononuclear cells and granulocytes from human blood. *Scand. J. Clin. Lab. Invest.* **21** (Suppl. 97): 77–89

Brown, J. (ed.) (1987) *Human Monoclonal Antibodies*. IRL Press, Oxford

Bruck, C., Portetelle, D., Glineur, C. and Bollen, A. (1982) One-step purification from ascitic fluid by DEAE Affi-Gel Blue chromatography. *J. Immunol. Methods* **53**: 313–319

Burchiel, S. W., Billman, J. R. and Alber, T. R. (1984) Rapid and efficient purification of mouse monoclonal antibodies from ascitic fluid using HPLC. *J. Immunol. Methods* **69**: 33–42.

Carson, D. A. and Freimark, B. D. (1986) Human lymphocyte hybridomas and monoclonal antibodies. *Adv. Immunol.* **38**: 275–311

Chen, T. R. (1977) In situ detection of mycoplasma contamination in cell cultures by fluorescent Hoechst 33258 stain. *Exp. Cell Res.* **104**: 255–262

Chi, D. S. and Harris, N. S. (1978) A simple method for the isolation of murine peripheral blood lymphocytes. *J. Immunol. Methods* **19**: 169–172

Childs, R. E. and Bardsley, W. G. (1975) The steady-state kinetics of peroxidase with 2,2′-azino-di-(3-ethyl-benzthiazoline-6-sulfonic acid) as chromogen. *Biochem. J.* **145**: 93

Chu, G. and Sharp, P. A. (1981) SV40 DNA transfection of cells in suspension: analysis of the efficiency of transcription and translation of T-antigen. *Gene* **13**: 197–202

Click, R. E., Benck, L. and Alter, B. J. (1972) Immune responses in vitro. I. Culture conditions for antibody synthesis. *Cell. Immunol.* **3**: 264–276

Coakley, W. T., Bardsley, D. W., Grundy, M. A., Zamani, F. and Clarke, D. J. (1989) Cell manipulation in ultrasonic standing wave fields. *J. Chem. Tech. Biotechnol.* **44**: 43–63

Coffino, P., Baumal, R., Laskov, R. and Scharff, M. D. (1972) Cloning of mouse myeloma cells and detection of rare variants. *J. Cell Physiol.* **79**: 429

Coligan, J. E., Gates, F. T., Kimball, E. S. and Maloy, W. L. (1983) Radiochemical sequence analysis of biosynthetically labelled proteins. *Methods Enzymol.* **91**: 413–434

Cook, W. D. and Scharff, M. D. (1977) Antigen-binding mutants of mouse myeloma cells. *PNAS* **74**: 5687–5691

Cote, R. J., Morrissey, D. M., Houghton, A. N., Beattie, E. J., Oettgen, H. F. and Old, L. J. (1983) Generation of human monoclonal antibodies reactive with cellular antigens. *PNAS* **80**: 2026–2030

Cotton, R. G. H. and Milstein, C. (1973) Fusion of two immunoglobulin-producing myeloma cells. *Nature* **244**: 42–43

Croce, C. M., Shander, M., Martinis, J., Cicurel, L., D'Ancona, G. G., Dolby, T. W. and Koprowski, H. (1979) Chromosomal location of the genes for human immunoglobulin heavy chains. *PNAS* **76**: 3416–3419

Croce, C. M., Linnenbach, A., Hall, W., Steplewski, Z. and Koprowski, H. (1980) Production of human hybridomas secreting antibodies to measles virus. *Nature* **288**: 488–489

Dangl, J. L. and Herzenberg, L. A. (1982) Selection of hybridomas and hybridoma variants using the fluorescence-activated cell sorter. *J. Immunol. Methods* **52**: 1–14

Dwyer, D. S., Bradley, J. R., Urguhart, C.K. and Kearney, J. F. (1983) Naturally occurring anti-idiotypic antibodies in myasthenia gravis patients. *Nature* **301**: 611–614

Edwards, P. A. W., Smith, C. M., Neville, A. M. and O'Hare, M. J. (1982) A human–human hybridoma system based on a fast-growing mutant of the ARH-77 plasma cell leukaemia-derived line. *Eur. J. Immunol.* **12**: 641–648

Ellens, D. J. and Gielkens, A. L. J. (1980) A simple method for the purification of 5-aminosalicylic acid. Application of the product as substrate in Enzyme Linked Immunosorbent Assay (ELISA). *J. Immunol. Methods* **37**: 325–332

Emanuel, D., Gold, J., Colacino, J., Lopez, C. and Hammerling, U. (1984) A human monoclonal antibody to cytomegalovirus (CMV). *J. Immunol.* **133**: 2202–2205

Engelman, E. G., Foung, S. K. H., Larrick, J. A. and Raubitschek, A. (eds) (1985) *Human Hybridomas and Monoclonal Antibodies.* Plenum Press, New York

Erikson, J., Martinis, J. and Croce, C. M. (1981) Assignment of the genes for human lambda immunoglobulin chains to chromosome 22. *Nature* **294**: 173–175

Erlanger, B. F. (1980) The preparation of antigenic hapten-carrier conjugates: A survey. *Methods Enzymol.* **70**: 85–104

Evans, H. J. and Vijayalaxmi, (1981) Induction of 8-azaguanine resistance and sister chromatid exchange in human lymphocytes exposed to mitomycin C and X rays in vitro. *Nature* **292**: 601–605

Ey, P. L., Prowse, S. J. and Jenkin, C. R. (1978) Isolation of pure IgG1, IgG2a and IgG2b immunoglobulins from mouse serum using protein-A Sepharose. *Immunochemistry* **15**: 429–436

Faulk, W. and Taylor, G. (1971) An immunocolloid method for the electronmicroscope. *Immunochemistry* **8**: 1081

Fazekas de St Groth, S. and Sheidegger, D. (1980) Production of monoclonal antibodies, strategy and tactics. *J. Immunol. Methods* **35**: 1–21

Festing, M. F. W., May, D., Connors, T. A., Lovell, D. and Sparrow, S. (1978) An athymic nude mutation in the rat. *Nature* **274**: 365–366

Flanagan, S. P. (1966) Nude: A new hairless gene with pleotropic effects in the mouse. Genetic Res. **8**: 295–309

Foung, S. K. H. and Perkins, S. (1989) Electric field-induced cell fusion and human monoclonal antibodies. *J. Immunol. Methods* **116**: 117–122

Fraker, P. J. and Speck, J. C. Jr (1978) Protein and cell membrane iodinations with a sparingly soluble chloroamide, 1, 3, 4, 6-tetrachloro-3a,6a-diphenyl glycouril. *Biochem. Biophys. Res. Commun.* **80**: 849–857

Freshney, R. I. (1987) *Culture of Animal Cells. A Manual of Basic Technique.* 2nd Edition. Alan R. Liss Inc., New York

Freund, Y. R. and Blair, P. B. (1982) Depression of natural killer activity and mitogen responsiveness in mice treated with Pristane. *J. Immunol.* **129**: 2826–2830

Galfre, G. and Milstein, C. (1981) Preparation of monoclonal antibodies: strategies and procedures. *Methods Enzymol.* **73**: 3–46

Galfre, G., Howe, S. C., Milstein, C., Butcher, G. W. and Howard, J. C. (1977) Antibodies to major histocompatibility antigens produced by hybrid cell lines. *Nature* **266**: 550–522

Galfre, G., Milstein, C. and Wright, B. (1979) Rat×Rat hybrid myelomas and a monoclonal anti-Fd portion of mouse IgG. *Nature* **277**: 131–133

Galfre, G., Cuello, A. C. and Milstein, C. (1981) New tools for immunochemistry: Internally labelled monoclonal antibodies, pp. 159–162. In *Monoclonal Antibodies and Developments in Immunoassay* (A. Albertini and R. Ekins, eds), Elsevier, Amsterdam

Gallyas, F., Gorcs, T., and Merchenthaler, I. (1982) High-grade intensification of the end-product of the diaminobenzidine reaction for peroxidase histochemistry. *J. Histochem. Cytochem.* **30**: 183–184

Gefter, M. L., Margulies, D. H. and Scharff, M. D. (1977) A simple method for polyethylene glycol promoted hybridisation of mouse myeloma cells. *Somatic Cell Genet.* **3**: 231

Gilliland, L. K., Clark, M. R. and Waldmann, H. (1988) Universal bispecific antibody for targeting tumor cells for destruction by cytotoxic T cells. *PNAS* **85**: 7719–7723

Giorgi, J. V., Burton, R. C., Barrett, L. V., Delmonico, S. L., Goldstein, G. and Cosimi, A. B. (1983) Immunosuppressive effect and immunogenicity of OKT11A monoclonal antibody in monkey allograft recipients. *Transplant. Proc.* **15**: 639–642

Glassy, M. C., Handley, H. H., Hagiwara, H. and Royston, I. (1983) UC729-6, a human lymphoblastoid B cell line useful for generating antibody secreting human–human hybridomas. *PNAS* **80**: 6327–6331

Goding, J. W. (1976a) The chromic chloride method of coupling antigens to erythrocytes: definition of some important parameters. *J. Immunol. Methods* **10**: 61

Goding, J. W. (1976b) Conjugation of antibodies with fluorochromes: modifications to the standard methods. *J. Immunol. Methods* **13**: 215–226

Goodfriend, T. L., Levine, L. and Fasman, G. D. (1964) Antibodies to bradykinin and angiotensin: A use of carbodiimides in immunology. *Science* **144**: 1344–1346

Graham, R. C., Jr and Karnovsky, M. J. (1966) The early stages of absorption of injected horseradish peroxidase in the proximal tubules of mouse kidney: Ultrastructural cytochemistry by a new technique. *J. Histochem. Cytochem.* **14**: 292–302

Graham, R. C., Lundholm, U. and Karnovsky, M. J. (1965) Cytochemical demonstration of the peroxidase activity with 3-amino-9-ethylcarbazole. *J. Histochem. Cytochem.* **13**: 150–152

Greenwood, F. C., Hunter, W. M. and Glover, J. S. (1963) The preparation of ^{131}I-labelled human growth hormone of high specific radioactivity. *Biochem. J.* **89**: 114–123

Groves, D. J., Morris, B. A., Tan, K., Da Silva, M. and Clayton, J. (1987) Production of an ovine monoclonal antibody to testosterone by an interspecies fusion. *Hybridoma* **6**: 71

Guesdon, J-L., Ternynck, T. and Avrameas, S. (1979) The use of avidin–biotin interaction in immunoenzymatic techniques. *J. Histochem. Cytochem.* **27**: 1131–1139

Guidry, A. J., Srikumaran, S. and Goldsby, R. A. (1986) Production and characterisation of bovine immunoglobulins from bovine×murine hybridomas. *Methods Enyzmol.* **121**: 244–265

Hale, G., Swirsky, D. M., Hayhoe, F. G. J. and Waldmann, H. (1983) *Mol. Biol. Med.* **1**: 321–334

Hamaguchi, Y., Yoshitake, S., Ishikawa, E., Endo, Y. and Ohtaki, S. (1979) Improved procedure for the conjugation of rabbit IgG and Fab' antibodies with β-D-galactosidase from Escherichia coli using N,N'-o-phenylenedimaleimide. *J. Biochem.* **85**: 1289–1300

Hatanaka, M., DelGiudice, R. A. and Long, C. (1975) Adenine formation from adenosine by mycoplasmas: Adenosine phosphorylase activity. *PNAS* **72**: 1401–1405

Hawkes, R., Niday, E. and Gordon, J. (1982) A dot-immunobinding assay for monoclonal and other antibodies. *Anal. Biochem.* **119**: 142–147

Herbert, W. J., Kristensen, F. with Aitken R. M., Eslami, M. B., Ferguson A., Gray, K. G. and Penhale, W. J. (1986) Laboratory animal techniques for immunology. In *Handbook of Experimental Immunology*, Vol. 4 (D. M. Weir, L. A. Herzenberg, C. Blackwell and L. A. Herzenberg, eds) chapter 133, Blackwell Scientific Publications, Oxford

Herzenberg, L. A. (1978) The fluorescence-activated cell sorter (FACS): A retrospective and prospective view. In *Immunofluorescence and Related Staining Techniques* (Knapp, Holubar and Wick, eds), Elsevier, Amsterdam

Hoffeld, J. T. (1981) Agents which block membrane lipid peroxidation enhance mouse spleen cell immune activities in vitro: relationship to the enhancing activity of 2-mercaptoethanol. *Eur. J. Immunol.* **11**: 371–376

Hoffeld, J. T. and Oppenheim, J. J. (1980) Enhancement of the primary antibody response by 2-mercaptoethanol is mediated by its action on glutathione in the serum. *Eur. J. Immunol.* **10**: 391–395

Holgate, C., Jackson, P., Cowen, P. and Bird, C. (1983) Immunogold–silver staining: New method of immunostaining with enhanced sensitivity. *J. Histochem. Cytochem.* **31**: 938

Honjo, T. (1983) Immunoglobulin genes. *Annu. Rev. Immunol.* **1**: 499–528

Houghton, A. N., Brooks, H., Cote, R. J., Taormina, M. C., Oettgen, H. F. and Old, L. J. (1983) Detection of cell surface and intracellular antigens by human monoclonal antibodies. *J. Exp. Med.* **158**: 53–65

Hubbard, A. L. and Cohn, Z. A. (1975) Externally disposed plasma membrane proteins 1. Enyzmatic iodination of mouse L cells. *J. Cell Biol.* **64**: 438–460

Institute of Animal Technology (IAT) Videos (address in Appendix C)

Ishikawa, E., Imagawa, M., Hashida, S., Yoshitake, S., Hamaguchi, Y. and Ueno, T. (1983) Enzyme-labelling of antibodies and their fragments for enyzme immunoassay and immunohistochemical staining. *J. Immunoassay* **4**: 209–327

James, K. and Bell, G. T. (1987) Human monoclonal antibody production. *J. Immunol. Methods* **100**: 5–40

Jasiewicz, M. L., Shoenberg, D. R. and Mueller, G. C. (1976) Selective retrieval of biotin-labelled cells using immobilized avidin. *Exp. Cell Res.* **100**: 213–217

Johnson, H. M., Brenner, K. and Hall, H. E. (1966) The use of water soluble carbodiimide as a coupling reagent in the passive haemagglutination test. *J. Immunol.* **97**: 171

Jones, P. T., Dear, P. H., Foote, J., Neuberger, M. S. and Winter, G. (1986) Replacing the complementarity-determining regions in a human antibody with those from a mouse. *Nature* **321**: 522–525

Juarez-Salinas, H., Engelhorn, S. C., Bigbee, W. L., Lowry, M. A. and Stanker, L. H. (1984) *Biotechniques* **2**: 164

Kabat, E. A., Wu, T. T., Reid Miller, M., Perry, H. M. and Gottesman, K. S. (1987) In *Sequences of Proteins of Immunological Interest*, US Department of Health and Human Services, US Government Printing Office

Kaplan, M. E. and Clark, C. (1974) An improved rosetting assay for the detection of human T lymphocytes. *J. Immunol. Methods* **5**: 131

Kato, K., Hamaguchi, Y., Fukui, H. and Ishikawa, E. (1975) Enzyme linked immunoassay II. A simple method for synthesis of the rabbit antibody-β-D-galactosidase complex and its general applicability. *J. Biochem.* **78**: 423–425

Kearney, J. F., Radbruch, A., Liesegang, B. and Rajewsky, K. (1979) A new mouse myeloma cell line that has lost immunoglobulin expression but permits the construction of antibody-secreting hybrid cell lines. *J. Immunol.* **123**: 1548–1550

Kelly, P. J., Millican, K. and Organ, P. (eds) (1988) *Principles of Animal Technology*. Institute of Animal Technology, Oxford

Kemshead, J. and Ugelstad, J. (1985) Magnetic separation techniques: Their application to medicine. *Mol. Cell. Biochem.* **67**: 11–18

Kendall, C., Ionescu Matiu, I. and Dreesman, G. R. (1983) Utilization of the biotin/avidin system to amplify the sensitivity of the enzyme linked immunosorbent assay (ELISA) *J. Immunol. Methods* **56**: 329–339

Kilmartin, J. V., Wright, B. and Milstein, C. (1982) Rat monoclonal antibodies derived by using a new non-secreting rat cell line. *J. Cell Biol.* **93**: 576–582

Kitagawa, T. and Aikawa, T. (1976) Enzyme coupled immunoassay of insulin using a novel coupling reagent. *J. Biochem.* **79**: 233–236

Klerx, J. P. A. M., Jansen Verplanke, C., Blonk, C. G. and Twaalhoven, L. C. (1988) In vitro production of monoclonal antibodies under serum free conditions using a compact and inexpensive hollow fibre cell culture unit. *J. Immunol. Methods* **111**: 179–188

Kohler, G. (1979) Soft agar cloning of lymphoid tumor lines: Detection of hybrid clones with anti SRBC activity. In *Immunological Methods*. Academic Press, pp. 397–401

Kohler, G. and Milstein, C. (1975) Continuous cultures of fused cells secreting antibody of predefined specificity. *Nature* **256**: 495–497

Kohler, G. and Milstein, C. (1976) Derivation of specific antibody-producing tissue culture and tumour lines by cell fusion. *Eur. J. Immunol.* **6**: 511–579

Kozbor, D. and Roder, J. C. (1981) Requirements for the establishment of high-titered human monoclonal antibodies against tetanus toxoid using the Epstein–Barr Virus technique. *J. Immunol.* **127**: 1275–1280

Kozbor, D. and Roder, J. C. (1983) The production of monoclonal antibodies from human lymphocytes. *Immunology Today* **4**: 72–79

Kozbor, D., Lagarde, A. E. and Roder, J. C. (1982) Human hybridomas constructed with antigen-specific Epstein–Barr Virus-transformed cell lines. *PNAS* **79**: 6651–6655

Kozbor, D., Tripputi, P., Roder, J. C. and Croce, C. M. (1984) A human hybrid myeloma for production of human monoclonal antibodies. *J. Immunol.* **133**: 3001–3005

Kozbor, D., Roder, J. C., Sierzega, M. E., Cole, S. P. C. and Croce, C. M. (1986) Comparative phenotypic analysis of available human hybridoma fusion partners. *Methods Enzymol.* **121**: 120–140

Kozbor, D., Burioni, R., Abramow Newerly, W., Roder, J. and Croce, C. M. (1987) The use of fusion partners in the production of human monoclonal antibodies. In *Human Monoclonal Antibodies: Current Techniques and Future Perspectives* (J. Brown, ed.), IRL Press Ltd, Oxford

Kronick, M. N. and Grossman, P. D. (1983) Immunoassay techniques with fluorescent phycobiliprotein conjugates. *Clin. Chem.* **29**: 1582–1586

Lamoyi, E. (1986) Preparation of F(ab')$_2$ fragments from mouse IgG of various subclasses. *Methods Enzymol.* **121**: 652–663

Lamoyi, E. and Nisonoff, A. (1983) Preparation of F(ab')$_2$ fragments from mouse IgG of various subclasses. *J. Immunol. Method.* **56**: 235

Lanzavecchia, A. and Scheidegger, D. (1987a) The use of hybrid hybridomas to target human cytotoxic T lymphocytes. *Eur. J. Immunol.* **17**: 105–111

Lanzavecchia, A. and Scheidegger, D. (1987b) The production of hybrid monoclonal antibodies capable of targeting human cytotoxic T cells. In *Human Monoclonal Antibodies* (J. Brown, ed.), IRL Press, Oxford

Lernhardt, W., Andersson, J., Courtinuho, A. and Melchers, F. (1978) Cloning of murine transformed cell lines in suspension culture with efficiencies near 100%. *Exp. Cell. Res.* **111**: 309–316

Lo, M. M. S., Tsong, T. Y., Conrad, M. K., Strittmatter, S. M., Hester, L. D. and Snyder, S. H. (1984) Monoclonal antibody production by receptor-media for electrically induced cell fusion. *Nature* **310**: 792–794

Marchalonis, J. J. (1969) An enzymic method for the trace iodination of immunoglobulins and other proteins. *Biochem. J.* **113**: 299–305

Markwell, M. A. K. (1982) A new solid state reagent to iodinate proteins. *Anal. Biochem.* **125**: 427–432

Mason, D. Y. and Sammons, R. E. (1978) Alkaline phosphatase and peroxidase for double immuno enzymatic labelling of cellular constituents. *J. Clin. Pathol.* **31**: 454–460

McGarrity, G. J. (1982) Detection of mycoplasmal infection of cell cultures. In *Advances in Cell Culture* Vol. 2, Academic Press, New York, pp. 99–131

McCarrity, G. J. and Kotani, H. (1985) Cell culture mycoplasmas. In *The Mycoplasmas*, Vol. IV, *Mycoplasma Pathogenicity* (Razin, A. & Barille, eds), Academic Press, New York.

McGarrity, G. J., Murphy, D. G. and Nichols, W. W. (eds) (1978) *Mycoplasma Infection of Cell Cultures.* Plenum, New York.

Metcalf, D. (1977) Hemopoietic colonies: In vitro cloning of normal and leukemic cells. *Recent Results Cancer Res.* **61**: 1–227

Miller, G., Shope, T., Lisco, H., Stitt, D. and Lipman, M. (1972) EBV: Transformation, cytopathic changes and viral antigens in squirrel monkey and marmoset leucocytes. *PNAS* **69**: 383

Milstein, C. (1986) From antibody structure to immunological diversification of the immune response. *Science* **231**: 1261–1269

Milstein, C. (1987) Diversity and the genesis of high affinity antibodies. *Biochem. Soc. Trans* **15**: 779–787

Milstein, C. and Cuello, A. C. (1983) Hybrid hybridomas and their use in immunohistochemistry. *Nature* **305**: 537–540

Mizobe, F., Martial, E., Colby Germinario, S. and Livett, B. G. (1982) An improved technique for the isolation of lymphocytes from small volumes of peripheral mouse blood. *J. Immunol. Methods* **48**: 269–279

Moeremans, M., Daneels, G., Van Dijck, A., Langanger, G. and DeMey, J. (1984) Sensitive visualisation of antigen–antibody reactions in dot and blot immune overlay assays with immunogold and immunogold/silver staining. *J. Immunol. Methods* **74**: 353–360

Morrison, S. L., Johnson, M. J., Herzenberg, L. A. and Oi, V. T. (1984) Chimeric human antibody molecules: Mouse antigen binding domains with human constant region domains. *Immunology* **81**: 6851–6855

Motta, G. and Locker, D. (1986) Detection of antibody-secreting hybridomas with diazobenzyloxymethyl paper. *Methods Enzymol* **121**: 491–497

Mowles, J. M. (1988) The use of ciprofloxacin for the elimination of mycoplasma from naturally infected cell lines. *Cytotechnology* **1**: 355–358

Nakane, P. K. (1968) Simultaneous localisation of multiple tissue antigens using the peroxidase labelled antibody method. *J. Histochem. Cytochem.* **16**: 557

Nakane, P. K. and Kawaoi, A. (1974) Peroxidase labelled antibody. A new method of conjugation. *J. Histochem. Cytochem.* **22**: 1084–1091

Neuberger, M. S. (1983) Expression and regulation of immunoglobulin heavy chain genes transfected into lymphoid cells. *EMBO J.* **2**: 1373–1378

Neuberger, M. S. and Rajewsky, K. (1981) Switch from hapten-specific immunoglobulin M to immunoglobulin D secretion in a hybrid mouse cell line. *PNAS* **78**: 1138–1142

Neuberger, M. S., Williams, G. T. and Fox, R. O. (1984) Recombinant antibodies possessing novel effector functions. *Nature* **312**: 604–608

Neuberger, M. S., Williams, G. T., Mitchell, E. B., Jouhal, S. S., Flanagan, J. G. and Rabbitts, T. H. (1985) A hapten-specific chimaeric IgE antibody with human physiological effector function. *Nature* **314**: 268–270

Nossal, G. J. V. and Pike, B. L. (1976) Single cell studies on the antibody-forming potential of fractionated, hapten-specific B lymphocytes. *Immunology* **30**: 189–202

Nyborg, W. L. (1977) Physical principles of ultrasounds. In *Ultrasound: Its Applications in Medicine and Biology* (F. J. Fry, ed.), Elsevier, pp. 1–76

Nygren, H. (1982) Conjugation of horseradish peroxidase to Fab fragments with different homobifunctional and heterobifunctional cross-linking reagents. *J. Histochem. Cytochem.* **30**: 407

Ochi, A., Hawley, R. G., Hawley, T., Schulman, M. J., Traunecker, A., Kohler, G. and Hozumi, N. (1983) Functional immunoglobulin M production after transfection of cloned immunoglobulin heavy and light chain genes into lymphoid cells. *PNAS* **80**: 6351–6355

O'Hare, M. J. and Yiu, C. Y. (1987) Human monoclonal antibodies as cellular and molecular probes: A review. *Molec. Cell. Probes* **1**: 33–54

Ohnishi, K., Chiba, J., Goto, Y. and Tokunaga, T. (1987) Improvement in the basic technology of electrofusion for generation of antibody-producing hybridomas. *J. Immunol. Methods* **100**: 181–189

Oi, V. T., Jones, P. P., Goding, J. W. and Herzenberg, L. A. (1978) Properties of monoclonal antibodies to mouse Ig allotypes, H-2 & Ia antigens. *Current Topics Microbiol. Immunol.* **81**: 115

Oi, V. T., Glazer, A. N. and Stryer, L. (1982) Fluorescent phycobiliprotein conjugates for analyses of cells and molecules. *J. Cell Biol.* **93**: 981–986

Oi, V. T., Morrison, S. L., Herzenberg, L. A. and Berg, P. (1983) Immunoglobulin gene expression in transformed lymphoid cells. *PNAS* **80**: 825–829

Olsson, L. and Kaplan, H. S. (1980) Human–human hybridomas producing monoclonal antibodies of predefined antigenic specificity. *PNAS* **77**: 5429–5431

Olsson, L., Kronstrom, H., Cambon-De Mouzon, A., Honsik, C., Brodin, T. and Jakobsen, B. (1983) Antibody producing human–human hybridomas. I. Technical aspects. *J. Immunol. Methods* **61**: 17–32

Orlandi, R., Gussow, D., Jones, P. T. and Winter, G. (1989) Cloning immunoglobulin variable domains for expression by the polymerase chain reaction. *PNAS* **86**: 3833–3837

Osband, M., Cavagnao, J. and Kupchick, H. Z. (1982) Successful production of human–human hybridoma IgG antibodies against Rh(D) antigen. *Blood* **60**(5), Suppl. 1, 81a (abstract)

O'Sullivan, M. J., Gnemmi, E., Chieregatti, G., Morris, D., Simmonds, A. D., Simmons, M., Bridges, J. W. and Marks, V. (1979) The influence of antigen properties on the conditions required to elute antibodies from immunoadsorbents. *J. Immunol. Methods* **30**: 127–137

Owens, C. S. (1981) High gradient magnetic separation of rosette forming cells. *Cell. Biophys.* **3**: 141

Owen, C. S., Winger, L. A., Symington, F. W. and Nowell, J. C. (1979) Rapid magnetic separation of rosette-forming lymphocytes. *J. Immunol.* **123**: 1778

Parham, P. (1983) On the fragmentation of monoclonal IgG1, IgG2a and IgG2b from Balb/c mice. *J. Immunol.* **131**: 2895–2902

Parks, D. R., Bryan, V. M., Oi, V. T. and Herzenberg, L. A. (1979) Antigen-specific identification and cloning of hybridomas with a Fluorescence Activated Cell Sorter. *PNAS* **76**: 1962

Paul, J. (1975) *Cell and Tissue Culture*. Churchill Livingstone, Edinburgh

Perez, A. G., Kim, J. H., Gelbard, A. S. and Djordjevic, B. (1972) Altered incorporation of nucleic acid precursors by mycoplasma-infected mammalian cells in culture. *Exp. Cell Res.* **70**: 301–310

Pontecorvo, G. (1975) Production of mammalian somatic cell hybrids by means of polyethylene glycol treatment. *Somatic Cell Genet.* **1**: 397–400

Pope, J. H., Scott, W. and Moss, D. J. (1974) Cell relationships in transformation of human leucocytes by Epstein–Barr Virus. *Int. J. Cancer* **14**: 122

Porstmann, B., Porstmann, T., Gaede, D., Nugel, E. and Egger, E. (1981) Comparison of chromogens for the determination of horseradish peroxidase as a marker in enzyme immunoassay. *J. Clin. Chem. Clin. Biochem.* **19**: 435–439

Potter, H., Weir, L. and Leder, P. (1984) Enhancer-dependent expression of human k immunoglobulin genes introduced into mouse pre B lymphocytes by electroporation. *PNAS* **81**: 7161–7165

Potter, M. and Boyce, C. (1962) Induction of plasma-cell neoplasms in strain Balb/c mice with mineral oil and mineral oil adjuvants. *Nature* **193**: 1086–1087

Potter, M. and Robertson, C. L. (1960) Development of plasma cell neoplasms in Balb/c mice after i.p. injection of paraffin-oil adjuvant, heat killed staphylococcus mixtures. *J. Natl Cancer Inst.* **25**: 847–862

Poupart, P., Vandenabeele, P., Cayphas, S., Van Snick, J., Haegeman, G., Kriegs, V., Fiers, W., and Content, J. (1987) B cell growth modulating and differentiating activity of recombinant human 26-KD protein (BSF-2, HuIFN-b2, HPGF). *EMBO J.* **6**: 1219

Radbruch, A., Liesegang, B. and Rajewsky, K. (1980) Isolation of variants of mouse myeloma X63 that express changed immunoglobulin class. *PNAS* **77**: 2909–2913

Rathlev, T. and Franks, G. F. (1982) New procedure for detecting anti-nuclear antibodies using glucose oxidase immunoenzyme technique *Am. J. Clin. Pathol.* **77**: 705–709

Raybould, T. J. G., Crouch, C. F., McDougall, L. J. and Watts, T. C. (1985a) Bovine–murine hybridoma that secretes bovine monoclonal antibody of defined specificity. *Am. J. Vet. Res.* **46**: 426

Raybould, T. J. G., Wilson, P. J., McDougall, L. J. and Watts, T. C. (1985b) A porcine–murine hybridoma that secretes porcine monoclonal antibody of defined specificity. *Am. J. Vet. Res.* **46**: 1768

Reading, C.L. (1982) Theory and methods for immunization in culture and monoclonal antibody production. *J. Immunol. Methods* **53**: 261–291

Reuveny, S., Velez, D., Miller, L. and Macmillan, J. D. (1986) Comparison of cell propagation methods for their effect on monoclonal antibody yield in fermentors. *J. Immunol. Methods* **86**: 61–69

Rice, D. and Baltimore, D. (1982) Regulated expression of an immunoglobulin k gene introduced into a mouse lymphoid cell line. *PNAS* **79**: 7862–7865

Riechmann, L., Clark, M., Waldmann, H. and Winter, G. (1988) Reshaping human antibodies for therapy. *Nature* **332**: 323–327

Rinderknecht, H. (1962) Ultrarapid fluorescent labelling of proteins. *Nature* **193**: 167–168

Roder, J. C., Cole, S. P. C. and Kozbor, D. (1986) The EBV hybridoma technique. *Methods Enzymol.* **121**: 140–167

Roitt, I., Brostoff, J. and Male, D. (1985) *Immunology*. Churchill Livingstone, Edinburgh

Rollin, B. E. (1987) Laws relevant to animal research in the United States. In *Laboratory Animals: An Introduction for New Experimenters* (A. A. Tuffery, ed.), J. Wiley & Sons Ltd, Chichester, pp. 323–333

Roth, J. (1983) The colloidal gold marker system for light and electron microscopic cytochemistry. *Techniques Immunochem.* **2**: 217

Rousseaux, J., Biserte, G. and Bazin, H. (1980) The differential enzyme sensitivity of rat immunoglobulin subclasses to papain and pepsin. *Mol. Immunol.* **17**: 469–482

Rousseaux, J., Pique, M. T., Bazin, H. and Biserte, G. (1981) Rat IgG subclasses: Differences in affinity to protein A-sepharose. *Mol. Immunol.* **18**: 639–645

Rousseaux, J., Rousseaux-Prevost, R. and Bazin, H. (1983) Optimal conditions for the preparation of Fab and F(ab')U2 fragments from monoclonal IgG of different rat IgG subclasses. *J. Immunol. Methods* **64**: 141–146

Rousseaux, J., Rousseaux-Prevost, R. and Bazin, H. (1986) Optimal conditions for the preparation of proteolytic fragments from monoclonal IgG of different rat IgG subclasses. *Methods Enzymol.* **121**: 663–669

Saiki, R. K., Scharf, S., Faloona, F., Mullis, K. B., Horn, G. T., Erlich, H. A. and Arnheim, N. (1985) Enzymatic amplification of β-globin genomic sequences and restriction site analysis for diagnosis of sickle cell anemia. *Science* **230**: 1350–1354

de Saint Vincent, B. R., Delbruck, S., Eckhart, W., Meinkoth, J., Vvitto, L. and Wahl, G. (1981) The cloning and reintroduction into animal cells of a functional CAD gene, a dominant amplifiable genetic marker. *Cell* **27**: 267–277

Sastry, L., Alting-Mees, M., Huse, W. D., Short, J. M., Sorge, J. A., Hay, B. N., Janda, K. D., Benkovic, S. J. and Lerner, R. A. (1989) Cloning of the immunological repertoire in Escherichia coli for generation of monoclonal catalytic antibodies: Construction of a heavy chain variable region-specific cDNA library. *PNAS* **86**: 5728–5732

Saul, F. A., Amzel, L. M. and Poljak, R. J. (1978) Preliminary refinement and structural analysis of the Fab fragment from human immunoglobulin NEW at 2.0 resolution. *J. Biol. Chem.* **253**: 585–597

Schimmelpfeng, L., Langenburg, U. and Hinrich Peters, J. (1980) Macrophages overcome mycoplasma infections of cells in vitro. *Nature* **285**: 661–662

Schmidt, J. and Erfle, V. (1984). Elimination of mycoplasmas from cell cultures and establishment of mycoplasma-free cell lines. *Exp. Cell Res.* **152**: 565–570

Schmitt, K., Daubener, W., Bitter-Suermann, D. and Hadding, U. (1988) A safe and efficient method for elimination of cell culture mycoplasmas using ciprofloxacin. *J. Immunol. Methods* **109**: 17–25

Schroeder, H. R., Boguslaski, R. C., Carrico, R. J. and Buckler, R. T. (1978) Monitoring specific protein binding reactions with chemiluminescence. *Methods Enzymol.* **57**: 424–425

Sharon, J., Morrison, S. L. and Kabat, E. A. (1980) Formation of hybridoma clones in soft agarose: Effect of pH and medium. *Somatic Cell Genet.* **6**: 435–441

Shulman, M., Wilde, C. D. and Kohler, G. (1978) A better cell line for making hybridomas secreting specific antibodies. *Nature* **276**: 269–270

Sofroniew, M. V. and Schrell, U. (1982) Long term storage and regular repeated use of diluted antisera in glass staining jars for increased sensitivity, reproducibility and convenience of single and 2-color light microscopic immunocytochemistry. *J. Histochem. Cytochem.* **30**: 504

Sompayrac, L. M. and Danna, K. J. (1981) Efficient infection of monkey cells with DNA of Simian Virus 40. *PNAS* **78**: 7575–7579

Stanbridge, E. (1971) Mycoplasmas and cell cultures. *Bacteriol. Rev.* **35**: 206

Stanworth, D. R. and Turner, M. W. (1978) Immunochemical analysis of immunoglobulins and their subunits. Chapter 6. In *Handbook of Experimental Immunology* (D. M. Weir, ed.), 3rd edition, Blackwell Scientific Publications, Oxford

Steinitz, M., Klein, G., Koskimies, S. and Makela, O. (1977) Epstein–Barr Virus induced B lymphocyte cell lines producing specific antibody. *Nature* **269**: 420–422

Strike, L. E., Devens, B. H. and Lundak, R. L. (1984) Production of human–human hybridomas secreting antibody to sheep erythrocytes after in vitro immunisation. *J. Immunol.* **132**: 1798–1803

Suresh, M. R., Cuello, A. C. and Milstein, C. (1986) Advantages of bispecific hybridomas in one-step immunocytochemistry and immunoassays. *PNAS* **83**: 7989

Tack, B. F., Dean, J., Eilat, D., Lorenz, P. E. and Schechter, A. N. (1980) Tritium labelling of proteins to high specific radioactivity by reduction methylation. *J. Biol. Chem.* **255**: 8842–8847

Teng, N. N. H., Lam, K. S., Reira, F. C. and Kaplan, H. S. (1983) Construction and testing of mouse–human heteromyelomas for human monoclonal antibody production. *PNAS* **80**: 7308

The, T. H. and Feltkamp, T. E. W. (1970a) Conjugation of fluoresceinisothiocyanate to antibodies. I. Experiments on the conditions of conjugation. *Immunology* **18**: 865–873

The, T. H. and Feltkamp, T. E. W. (1970b) Conjugation of fluoresceinisothiocyanate to antibodies. II. A reproducible method. *Immunology* **18**: 875–881

Thorell, J. I. and Johansson, B. G. (1971) Enzymatic iodination of polypeptides with ^{125}I to high specific activity. *BBA* **251**: 363–369

Thorpe, G. H. G., Williams, L. A., Kricka, L. J., Whitehead, T. P., Evans, H. and Stanworth, D. R. (1985) A rapid luminescently monitored enzyme immunoassay for IgE. *J. Immunol. Methods* **79**: 57

Tijssen, P. (1985) *Practice and Theory of Enzyme Immunoassays*. Vol. 15 of *Laboratory Techniques in Biochemistry and Molecular Biology* (Burdon, R. H. & van Knippenberg, P. H. eds), Elsevier Science Publishers, Amsterdam

Tubbs, R. R. and Shebani, K. (1981) Chromogens for immunohistochemistry. *J. Histochem. Cytochem.* **29**: 684

Tucker, E. M., Dain, A. R., Wright, L. J. and Clarke, S. W. (1984) Specific bovine monoclonal antibodies produced by a re-fused mouse/calf hybridoma. *Hybridoma* **3**: 171–176

Tuffery, A. A. (ed.) (1987) *Laboratory Animals. An Introduction for New Experimenters.* J. Wiley & Sons Ltd.

University Federation for Animal Welfare (UFAW) (1986) *Handbook on the Care and Management of Laboratory Animals.* Longmans, London and New York

US Department of Health and Human Services (1985) *Guide for the Care and Use of Laboratory Animals.* National Institutes of Health, Bethesda, Maryland

Van Meel, F. C. M., Steenbakkers, P. G. A. and Oomen, J. C. H. (1985) Human and chimpanzee monoclonal antibodies. *J. Immunol. Methods* **80**: 267

Van Meurs, G. J. E. and Jonker, M. (1986) Production of primate monoclonal antibodies *J. Immunol. Methods* **95**: 123

Van Snick, J., Cayphas, S., Vink, A., Uytenhove, C., Coulie, P. G., Rubira, M. R. and Simpson, R. J. (1986) Purification and NHU2-terminal amino acid sequence of a T-cell derived lymphokine with growth factor activity for B-cell hybridomas. *PNAS* **83**: 9679

Verhoeyen, M., Milstein, C. and Winter, G. (1988) Reshaping human antibodies: Grafting an antilysozyme activity. *Science* **239**: 1534–1536

Vienken, J. and Zimmermann, U. (1985a) An improved electrofusion technique for production of mouse hybridoma cells. *FEBS* **182**(2): 278–280

Vienken, J., Zimmermann, U., Zenner, H. P., Coakley, W. T. and Gould, R. K. (1985b) Electroacoustic fusion of erythrocytes and of myeloma cells. *Biochim Biophys. Acta* **820**: 259–264

Vitetta, E. S., Capra, J. D., Klapper, D. G., Klein, J. and Uhr, J. W. (1976) The partial amino sequence of an H-2K molecule. *PNAS* **73**: 905–909

Voller, A., Bidwell, D. E. and Bartlett, A. (1979) *The enzyme linked immunosorbent assay.* Dynatech (see Appendix C)

Walker, S. M., Meinke, G. C. and Weigle, W. O. (1977) Enrichment of antigen specific B lymphocytes by the direct removal of B cells not bearing specificity for the antigen. *J. Exp. Med.* **146**: 445–456

Walker, S. M., Meinke, G. C. and Weigle, W. O. (1979) Separation of various B-cell subpopulations from mouse spleen. 1) Depletion of B-cells by rosetting with glutaraldehyde-fixed anti-immunoglobulin coupled red blood cells. *Cell Immunol.* **46**: 158

Wall, R. and Kuehl, M. (1983) Biosynthesis and regulation of immunoglobulins. *Annu. Rev. Immunol.* **1**: 393–422

Walls, E. V., Lam, K. and Crawford, D. H. (1988) Production of human monoclonal antibodies using EBV. pp. 21–32. In *Clinical Applications of Monoclonal Antibodies* (Hubbard, R. and Marks, V., eds), Plenum Pub. Corp., New York

Ward, E. S., Gussow, D., Griffiths, A. D., Jones, P. T. and Winter, G. (1989) Binding activities of a repertoire of single immunoglobulin variable domains secreted from Escherichia coli. *Nature* **341**: 544–546

Weeks, I. and Woodhead, J. S. (1984) Chemiluminescence immunoassay. *J. Clin. Immunoassay* **7**: 82–89

Weeks, I., Beheshti, I., McCapra, F., Campbell, A. K. and Woodhead, J. S. (1983) Acridinium esters as high specific activity labels in immunoassay. *Clin. Chem.* **29**: 1474–1479

Weir, D. M. (1986) *Handbook of Experimental Immunology.* Blackwell Scientific Publications, Oxford.

Wells, D. E. and Price, P. J. (1983) Simple rapid methods for freezing hybridomas in 96 well microculture plates. *J. Immunol. Methods* **59**: 49–52

Wilcheck, M. and Bayer, E. A. (1984) *Immunology Today* **5**: 39

Wilson, M. B. and Nakane, P. K. (1978) In *Immunofluorescence and Related Techniques* (W. Knapp, H. Holubar and G. Wick, eds), Elsevier, Amsterdam

Wojchowski, D. M. and Sytkowski, A. J. (1986) Hybridoma production by simplified avidin-mediated electrofusion. *J. Immunol. Methods* **90**: 173–177

Woodhead, J. S., Simpson, J. S. A., Weeks, I., Patel, A., Campbell, A. K., Hart, R., Richardson, A. and McCapra, F. (1981) Chemiluminescent labelled antibody techniques. In *Monoclonal Antibodies and Developments in Immunoassay* (A. Albertini and R. P. Ekins, eds), Elsevier, Amsterdam pp. 135–145

Yarmush, M. L., Gates III, F. T., Weisfogel, D. R. and Kindt, T. J. (1980) Identification and characterisation of rabbit–mouse hybridomas secreting rabbit immunoglobulin chains. *PNAS* **77**: 2899

Yoshitake, S., Imagawa, M., Ishikawa, E., Niitsu, Y., Urushizaki, I., Nishiura, M., Kanazawa, R., Kurosaki, H., Tachibana, S., Nakazawa, N. and Ogawa, H. (1982) Mild and efficient conjugation of rabbit Fab′ and horseradish peroxidase using a maleimide compound and its use for enzyme immunoassay. *J. Biochem.* **92**: 1413

Zimmermann, U. (1982) Electric field-mediated fusion and related electrical phenomena. *Biochim. Biophys. Acta* **694**: 227–277

Zimmermann, U. (1986) Electrical breakdown, electropermeabilisation and electrofusion. *Rev. Physiol. Biochem. Pharmacol.* **105**: 175

Suggestions for further reading

DATABASES

SUBIS (Sheffield University Biomedical Information Service) Bulletin on Monoclonal Antibodies (Address in Appendix C).
CABS (Current Awareness in Biological Sciences) Database: Current Advances in Immunology (Address in Appendix C).

TECHNICAL BOOKS

Monoclonal Antibody Technology. A. M. Campbell (1984) Elsevier (Laboratory Techniques Series), Amsterdam.
Monoclonal Antibodies: Principles and Practice. James W. Goding (1983, 1986) Academic Press, London.

APPLICATIONS

Antibody as a Tool: The Applications of Immunochemistry (1982) Edited by J. J. Marchalonis and G. W. Warr, John Wiley & Sons, Chichester.
Monoclonal Antibodies. K. Sikora and H. M. Smedley (1984) Blackwell Scientific Publications, Oxford
Clinical Applications of Monoclonal Antibodies. British Medical Bulletin (1984) Vol. 40, No. 3, Editor E. E. Lennox. Published for the British Council by Churchill Livingstone, London.
Monoclonal Antibodies: Probes for the Study of Autoimmunity and Immunodeficiency. Edited by B. F. Haynes and G. S. Eisenbarth. Academic Press, London.

Monoclonal Antibodies to Receptors: Probes for Receptor Structure and Function. Edited by M. Greaves. Receptor and Recognition Series B, Vol. 17. Chapman and Hall, London.

Pregnancy Testing: Applications of Monoclonal Technology. H. Van Hell and J. Helmich. International Biotechnology Laboratory March/April 1984. pp. 22–33.

Monoclonal Antibodies in Clinical Medicine. A. J. McMichael and J. W. Fabre (1982) Academic Press, London.

Hybridomas in Cancer Diagnosis and Therapy. M. S. Mitchell and H. F. Oettgeu (1982) Raven Press, New York.

Monoclonal Antibodies for Diagnosis of Infectious Diseases in Humans. R. C. Nowinski et al (1982) *Science* **219**: 637–644.

The Use of Monoclonal Antibody Techniques in the Study of Developing Cell Surfaces. C. Milstein and E. Lennox (1980) *Current Topics Dev. Biol.* **14**: 1–32.

Clinical Potential of Monoclonal Antibodies. K. Sikora and H. M. Smedley (1982) *Cancer Surveys* **1**: 521–541.

Monoclonal Antibodies and Developments in Immunoassay. Edited by Alberto Albertini and Roger Ekins (1981) Elsevier, Amsterdam.

Monoclonal Antibodies in Plant Disease Research. E. L. Halk and S. H. DeBoer (1985) *Annu. Rev. Phytopathol.* **23**: 1321.

Monoclonal Antibodies against Bacteria. Edited by A. J. L. Marcario and E. Conway de Marcario (1985) Academic Press, London.

Monoclonal Antibodies (Methods in Haematology, Vol. 13). Edited by P. C. L. Beverley (1986) Churchill Livingstone, Edinburgh.

Index

Radioimmunoassay (RIA)   63
Radiolabelling of antibodies   122
  Bolton–Hunter method   124
  chloramine T method   122
  iodogen method   124
  lactoperoxidase method   123
Rosetting of T-cells   81
RPMI 1640 medium, recipe   151

Saponin   40
Salvage pathway   68
Sendai virus in cell fusion   85
Scatchard analysis   111
Screening tests for antibody   48
  choice of   52
  design of   49
  when   48
Serum, see Foetal bovine serum
Single domain antibodies (dABs)   149
SMPB   128
SPDP   128
Spleen cells, preparation of   79
Sheep red blood cells (SRBC)   82
Sterilisation
  of glassware and dissecting instruments,
      16, 21
  of solutions   16
Streptomycin, recipe   152

6-Thioguanine (6-TG)   69
Titre determination   106

Tissue culture
  culture media
    choice of   22
    detailed recipes   151
    preparation of   23
  controlling infection   26
  general procedures   20
  maintenance of laboratory   19
  personal hygiene   20
  setting up a laboratory   8
  washing and sterilising   21

T-cell depletion of peripheral lymphocytes   81
Thioguanine   69, 154
Thymidine kinase (TK)   69, 146
Thymocytes, as feeder cells   94
TMB as HRP substrate   129, 157
Transfection   146
TRITC, see Fluorescent labelling
Tritium labelling of antibody   126, 138
Trypan blue   76
Trypsin for cell detachment   155

Urease   131
  conjugation method   132

Vectors   144

XGPRT   146

Yeast (infection in tissue culture)   27